**BioProtopia**
Designing the Built Environment
with Living Organisms

# BioProtopia
## Designing the Built Environment with Living Organisms


**Edited by**
Ruth Morrow
Ben Bridgens
Louise Mackenzie



Birkhäuser
Basel

# Glossary

# A–Z

## A

**adhesion**
↗ Page 158
The action or process of sticking to a surface object.

**anaerobic digestion**
↗ Page 133
The decomposition of organic matter by microorganisms in the absence of oxygen.

## B

**bacterial cellulose**
↗ Page 62
A biodegradable, natu cellulose that is produc by bacteria.

**biocement**
↗ Page 151
A generic name for the product that results fro the process of strengt ening soil through bio- mineralisation or micro bially induced calcium carbonate precipitatio

**biocompatibility**
↗ Page 63
Related to the behavio of biomaterials in vari- ous contexts. The term refers to the ability of a living material to perfo within a specific situa- tion (typically the huma body), without produci an adverse effect.

**biocompatible**
↗ Page 78
Not harmful to living ce

**biocomposite**
↗ Page 154
A material made up of two or more biomateri

**embodied carbon**
↗ Page 22
The $CO_2$ generated by goods, such as food or construction materials, throughout the life cycle of the goods. For example, the embodied carbon in a brick would include $CO_2$ emissions at every stage from raw material extraction, processing, transportation, construction, demolition and disposal (or reuse).

**epifluorescence microscopy**
↗ Page 126
An optical imaging technique that uses fluorescence instead of, or in addition to, scattering, reflection, and attenuation or absorption, to study the properties of organic or inorganic substances.

**eutrophication**
↗ Page 133
A chain reaction set off by increased concentrations of nutrients in a water body. Increased plant and algae growth eventually results in mass decomposition, generating excess $CO_2$ in the water. This lowers the water's pH, resulting in acidification and harming aquatic life.

## F

**feedstock**
↗ Page 143
In biotechnology, the raw biological material for an industrial or technological process. Normally converted to something different.

**fermentation**
↗ Page 64
The breakdown of chemical components in a medium by microorganisms, such as bacteria and yeasts, typically involving conversion into sugars or alcohol and often resulting in carbonation or heat.

## H

**HBBE**
↗ Page 19
The Hub for Biotechnology in the Built Environment – a multidisciplinary research hub which aims to create built environments which are life-sustaining and sustained by life (www.bbe.ac.uk).

**hydrogel**
↗ Page 164
A highly absorbent, water-based polymer that does not dissolve in water.

**hydrolysis**
↗ Page 105
The chemical reaction of an organic molecule with water to form two or more new substances.

**hygromorphs**
↗ Page 120
Materials and objects that respond to environmental humidity by changing their shape.

## K

**kappa-carrageenan**
↗ Page 164
A natural ingredient from red seaweed used as an additive to form strong, rigid gels.

**Kombucha Method**
↗ Page 140
The process by which a nutrient solution of tea and sugar is fermented, generating kombucha liquid and SCOBY. Yeast ferments sugar to alcohol, followed by the bacterial fermentation of alcohol to acetic acid.

# L

**lignin**
↗ Page 177
A polymer made up of aromatic subunits which is deposited on the cell walls of plants – resulting in the cells becoming rigid or woody.

**living construction**
↗ Page 118
The use of living cells to construct and transform materials, through methods of fabrication and assembly at multiple scales. The term also refers to how designers and architects employ these processes.

**longitudinal analysis**
↗ Page 98
A study in which the same data is collected at different periods, to observe changes over time.

# M

**metabolite**
↗ Page 178
A molecule produced by a reaction within a metabolic network.

**Microbial Fuel Cell (MFC)**
↗ Page 131
A bioelectrochemical device which uses bacteria to generate electricity from chemical energy in organic matter.

**Microbially Induced Calcium Carbonate Precipitation (MICCP or MICP)**
↗ Page 151
The process by which, under suitable environmental conditions, bacteria initiate the production of calcium carbonate.

**microbiome**
↗ Page 101
All of the microorganisms in a particular environment.

**more-than-human**
↗ Page 38
A term first used by David Abram in his book A Spell of the Sensuous (1996) to refer to the interconnected relationship between humans and the wider environment. Whilst the term can be used to refer to phenomena on a planetary scale, such as global warming, it can also refer to the composition of the human body, where microbial cells outnumber human cells by ten to one.

**morphology**
↗ Page 157
A particular form, shape or structure. In biology, morphology refers to the form (size, shape, structure and relationship of the constituent parts) of a biological organism, which may vary according to phenotype and environmental factors.

**mycelium**
↗ Page 62
The root-like, branching network of fungi commonly found on soil or other substrates.

# N

**net-zero**
↗ Page 92
Also referred to as 'carbon neutrality', the term means balancing $CO_2$ emissions with $CO_2$ removal to ensure that atmospheric $CO_2$ levels do not increase.

**non-human**
↗ Page 17
A term that refers to entities that are not human, but may be considered to have agency. The term is commonly used to describe living organisms that are not human but can also be applied to the environment, machines, objects, weather, and other nonliving things (see, for example, The Nonhuman Turn, edited by Richard Grusin, 2015). Within the context of this book, the term 'non-human' refers to living organisms.

**nucleation**
↗ Page 105
The beginning of the formation of a new structure via self-assembly or self-organisation.

**nutrient media**
↗ Page 154
Liquid or solid media containing the nutrients needed to support the growth of microorganisms.

# P

**parametric**
↗ Page 172
Parametric design is a method where the properties of elements (such as their geometry and form) are controlled by algorithms, in contrast to being designed directly. The role of the designer in a parametric design process is to define the controlling parameters and their relationships.

**petrographic thin sections**
↗ Page 93
The laboratory preparation of a geological sample, such as rock, bone, pottery, soil or metal, for analysis. The sample is prepared as a thin slice, typically around 30 microns (0.03 mm) thick.

**phyllotaxis pattern**
↗ Page 108
Spiralling patterns that arise from intersecting regular intervals in polar coordinate geometry – found in botanical structures such as pinecones and aloes.

**polymer**
↗ Page 63
A substance with a molecular structure that is composed of multiple, similar units bonded together.

**polymerisation**
↗ Page 78
A chemical reaction in which small molecules called monomers join together to make a large chain molecule called a polymer. The degree

of polymerisation has a dramatic effect on the mechanical properties of a polymer.

**polymorph**
↗ Page 94
In chemistry and geology a polymorph is any substance or mineral that exists in more than one form or crystal structure.

**probiotic**
↗ Page 128
Microbial cells (or a substance containing microbial cells) that are reported to have beneficial properties for the health and wellbeing of the host organism.

**prototype**
↗ Page 77
An early-stage trial of technologies, materials, processes and/or assemblies.

# R

**reaction-diffusion (Turing) pattern**
↗ Page 108
Spatially periodic patterns that arise from random or uniformly homogeneous distributions of two diffusible domains interacting with each other.

**rebound effect**
↗ Page 21
The rebound effect is the reduction in gains from new technologies which increase efficiency of resource use, due to behavioural changes.

**regenerative design**
↗ Page 22
A design approach which incorporates systems and processes which restore or replenish sources of materials and energy, creating systems which are self-sustaining and beneficial to society and to nature.

# S

**Scanning Electron Microscopy (SEM)**
↗ Page 126
A microscopy technique that produces images by scanning the surface of a sample with a focused beam of electrons. The electrons interact with atoms in the sample, producing various signals that contain information about the surface topography and composition of the sample.

**SCOBY**
↗ Page 58
Acronym for 'Symbiotic Culture Of Bacteria and Yeast'.

**shortest walk branching pattern**
↗ Page 108
A pattern with a generative path, connecting closest defined spatial nodes that split at given intervals as they approach a destination point.

**substrate**
↗ Page 62
A supporting or underlying substance. In biology, a substrate is the material upon which an organism resides, feeds and/or grows.

**sustainability**
↗ Page 101
Sustainable development meets the needs of the present without compromising the ability of future generations to meet their own needs.

# T

**the OME**
↗ Page 19
An experimental building designed for prototyping architectural applications of biotechnology in a domestic context. See Chapter 1.1 for further details.

# U

**UK Building Regulations**
↗ Page 22
Statutory regulations that ensure that buildings in the UK are safe and efficient to operate. In addition to the legal requirements, the Building Regulations also provide details of standard forms of construction which meet the regulations.

**urease**
↗ Page 105
An enzyme that catalyses the breakdown of urea via the process of hydrolysis, to form ammonia and carbon dioxide.

# V

**Voronoi pattern**
↗ Page 108
In biology, patterns with a tessellation of regions where all points within each region are closest to a randomly scattered reference point.

# X

**X-Ray Diffraction (XRD)**
↗ Page 94
A non-destructive technique for analysing the atomic or molecular structure of materials.

**X-ray microtomography**
↗ Page 93
A non-destructive imaging process that uses X-rays to create cross-sections of a physical object that can be used to create a virtual 3D visualisation.

# Contents

# Bioprotopia
## Reimagining the Lived Environment

↗ **Glossary**
prototype
non-human
microbiome
biomaterials

What might be called to mind by the term *bioprotopia*? The prefix 'bio-' comes from the Greek *bios*, or life, denoting biological processes and living organisms; 'proto-' indicates the earliest or original form suggesting new developments and **prototypes**↗, whilst also hinting at something ancient. The suffix '-topia' stems from *topos*, or place. In contemporary usage, it becomes the lived physical place, as opposed to the imagined space of a utopia, or paradise. Literally, bioprotopia translates as 'life, the earliest form of place'. This is the provocation suggested throughout this book: that place is not primarily associated with a *built* environment, but instead has always existed as the living environment. **Bioprotopia** suggests a new approach to understanding, making and defining place; that is, as something formed through our coexistence with living organisms and biological processes.

**Bioprotopia** offers a vision of buildings that can grow, self-heal and create virtuous cycles where the waste from one process feeds another: a vision where the spaces that we inhabit are attuned to both human occupants and **non-human**↗ microbial ecologies.

This is the first book to ground the concept of biotechnology in the built environment in tangible, large-scale outcomes. This visually rich book introduces the reader to biomaterials and bioprocesses that bring to life the diverse possibilities of designing, or rather evolving, environments with microorganisms.

**Bioprotopia** presents recent and ongoing research at the Hub for Biotechnology in the Built Environment (HBBE). The HBBE is a research collaboration funded by Research England which brings together bioscientists from Northumbria University (UK) and architects, designers and engineers from Newcastle University (UK) to work together towards a common vision: to make built environments which are life-sustaining and sustained by life. To achieve this, the HBBE is developing biotechnologies to create a new

**Figure 1**
Biosandstone cladding

generation of living buildings which are responsible and responsive to their natural environment. Buildings that are grown using living engineered materials to eliminate inefficient industrial construction processes; which metabolise their own waste to reduce pollution and generate energy, and modulate their **microbiome**↗ to benefit human and ecological health and wellbeing. The research is underpinned by a reflexive, inclusive and critical approach to biotechnology, embedded within the interactions that emerge across disciplines, sectors, technologies and species | **Figure 2**↗ |.

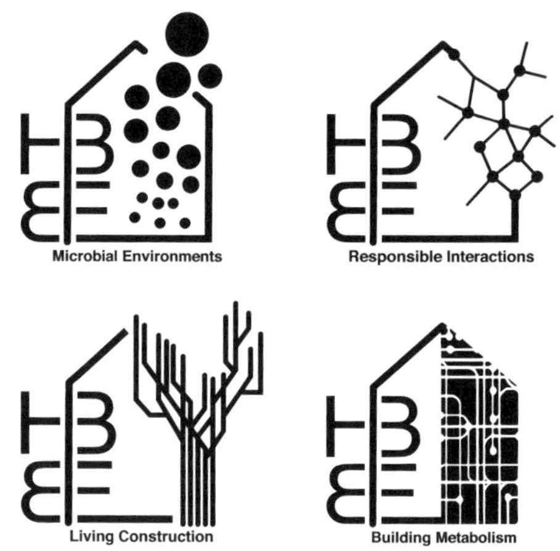


**Figure 2**
The logos of the four HBBE Research Themes

At the centre of the book sits a series of prototypes operating at the scale of the building. These 'MacroArchitectures' demonstrate the variety of creative, practical applications of biotechnology in the built environment. The MacroArchitectures test, for example, the potential of bacterial cellulose as a building material; the structural futures offered when three-dimensional knitted forms are impregnated with mycelium (fungi); concrete made from waste that hosts life; lime render that heals itself through the action of bacteria; systems for transforming household waste into energy, and systems for testing materials, scientifically and culturally, as hosts for microbial ecologies. Alongside the MacroArchitectures are a series of 'MicroAssemblies' – smaller-scale, early biotechnological prototypes. These suggest futures in which we cede control of form and aesthetics to the organisms that build our materials for us, or where the building materials that we design mimic natural processes and forms: where microbes within our buildings absorb carbon and where bacteria within our waste streams create new materials within our homes.

Not only does **Bioprotopia** propose potential applications of microbially led technologies, discussing openly the future scientific and technical challenges that require further resolution, but it also begins to outline the cultural shifts and emerging practices that will be required to realise a new era of **biomaterials**↗ and biosystems. This will fundamentally change how we build, operate and live. By locating these biotechnological architectures outside of the context of the laboratory, it draws attention to the direct experience of working with biomaterials and biosystems – the feel, the look, and the smell – and how these interactions profoundly challenge societal perceptions of what should and shouldn't constitute the built environment.

The book asks architects to abandon resource-intensive aesthetics: the smooth perfection of mass-produced, globally traded, environmentally destructive materials. Instead it radically shifts the focus to allow biomaterials and biosystems to develop their own lively architectural expression. At this early stage in the development of these technologies, the designer's role is not to predetermine the form or application of these living systems but rather to respond to, and creatively improvise with them. The biodesigner is positioned as someone who seeks to inform a wider social aesthetic by making the invisible visible.

Finally, the book makes clear that we are still at the start of these processes. In order to respond in a timely way and at a pace fitting to the climate emergency, not only do we urgently need to build interdisciplinary research teams, but we also need to work with the construction industry, governments and regulatory bodies, as well as wider DIY biodesign communities. Only by doing so can we collaborate effectively to address the challenges and opportunities that biotechnology offers. The greatest challenges do not lie in the technical application of these technologies, but rather in the social and cultural practices that surround them.

The essays and prototypes within this book present work that navigates across practices of architecture, design and biotechnology to explore what is possible and what is necessary when considering life as a form of place. The practical examples offered suggest biological forms of place within contemporary society. They seek to question the potential pitfalls that such a bioprotopia might encounter. Thus, the prototypes within these pages do not manifest either as utopia or dystopia, but as growing constructs: places made with life that therefore must evolve.

Ruth Morrow
Ben Bridgens
Louise Mackenzie

# 1

# Growing Space, Growing Discourse
## Developing a Language for Biological Architecture

↗ **Glossary**
sustainability
prototype
non-human

How do we define the spaces that we call our places of home, work and leisure, if not as the 'built environment'? Within architecture, the language for spaces that might be grown rather than built is yet to be developed. The chapters within this section move the reader towards thinking about what must be considered and what must change. Rather than describing a utopia in which we cohabit with other species in a perfect cycle of **sustainability**↗, these essays take into account the global infrastructures and contexts of the built environment that make working with biotechnology complex. Chapters hint at the likelihood of practical adoption across geographical boundaries, the complexities of moving from the lab into the environment, and the socio-political and cultural contexts in which these new biotechnological systems and processes will be received.

Chapter 1.1
↗Page 18

**Chapter 1.1** introduces us to the language of restoration and regeneration as architectural technologist Ben Bridgens likens future building methodologies to natural growth cycles, and suggests we require a paradigm shift in our approach to construction. In a frank discussion about the process of building an experimental building on a city-centre university campus in the UK, we begin to see the entrenched difficulties in bringing about a cultural shift in the long-established construction industry, built on the shoulders of a deeply problematic colonial and extractivist industrial past. In navigating the collaborative prototyping approach that shapes the material content of this book, the words 'adaptability' and 'anticipation' then become an important part of the architectural lexicon. Indeed these words extend their value beyond the **prototypes**↗ within these pages to the longer-term project of designing and constructing with living materials.

Chapter 1.2
↗Page 34

In **Chapter 1.2**, architect Ruth Morrow and artist Louise Mackenzie explore the possibility of moving towards a biological architecture by identifying concepts that help to define the field. We learn of researchers growing and tending to their materials, observing their changing states across seasons, nurturing them as they live and repurposing them when they die. Space and time are reconceptualised as necessarily slow and rhythmic as we begin to experience the concepts of 'making with' and 'living with' rather than 'using' materials and 'living in' buildings. Ideas of adaptability are echoed through a language of fluidity, as materials behave unexpectedly and we learn to be surprised and guided by their changing states. This in turn suggests a need to form new ethical and aesthetic relationships that embrace the moist and the lumpy. An argument emerges for two approaches: one where we embrace seasonal variation and weathering, and another where materials can be upgraded regularly as long as they are compostable – a 'fast-fashion' biological architecture. As we consider our spaces of dwelling to be multi-species habitats, we move towards the language of a *lived* rather than a *built* environment.

Chapter 1.3
↗Page 46

The final chapter in this section, **Chapter 1.3**, brings attention to the ethical relations that architecture must form in considering the social and environmental contexts of the spaces we inhabit. Drawing on two recent examples, the Grenfell Tower fire in the UK and the COVID-19 pandemic, architectural historian Peg Rawes highlights the need for care to become embedded in the language of architecture. Focusing on the biopolitical context in which architecture operates, Rawes introduces us to the concepts of 'kind matters' and 'matters of kind'; alluding in the first instance to the care that architects must take in considering the wider social, political and cultural contexts in which their designs are formed; and in the second, the importance of attending to *all who matter* within these spheres.

Through these chapters therefore, we seed the discourse necessary for the growth of a new language. A language which helps us consider the environments that we inhabit at the scale of both the planetary and the local. A language that reminds us that these environments are home to a variety of lives, both human and **non-human**↗. A language in which materials and processes are no longer static, but are in a constant state of transformation, and in which a collective 'we' is allowed space to grow.

# 1.1

Ben Bridgens

# 1.1

## Building Practice
## Looking Beyond 'Net-Zero' to Regenerative Architecture

↗ **Glossary**
sustainability
net-zero
the OME
Hub for Biotechnology in the Built Environment (HBBE)

This essay explores the possibility of going beyond **sustainability** ↗ and **net-zero** ↗ by working with biological processes and systems to enable us to construct, operate and maintain the built environment whilst restoring a diverse, abundant natural world. Two distinct challenges of instigating this profound shift are explored. The first is the difficulty in developing and testing new construction technologies which require multi-scale facilities and transgress disciplinary boundaries: architectural technology research facilities rarely include the infrastructure or expertise to carry out biological experimentation. The second is the challenge of introducing not only new materials, but entirely new forms of construction and building systems, into a stringently regulated and risk-averse construction industry. The construction of an experimental building at Newcastle University, called **the OME** ↗, is used as a case study to examine these challenges. The OME is the **Hub for Biotechnology in the Built Environment's (HBBE)** ↗ experimental building designed to promote collaboration across disciplines, provide a test bed for architectural scale prototyping of biological building technologies, and enable engagement with the public and industry | **Figure 1** ↗ |.

## Beyond Incremental Reductions in Environmental Impact

↗ **Glossary**
biodiversity
sustainability
net-zero
rebound effect
ecosystem
regenerative design

Fifty years after the disastrous implications of humankind's use of fossil fuels and other non-renewable resources was made clear (Meadows et al., 1972), there is finally a growing urgency to address the problem. We see all too clearly the impacts of climate change that include wholesale pollution and destruction of the natural world, wildfires, flooding, heatwaves and catastrophic loss of **biodiversity** ↗.

**'Sustainable development is development that meets the needs of the present without compromising the ability of future generations to meet their own needs'.** (Brundtland, 1987)

**Sustainability** ↗ is a complex, multi-faceted concept incorporating social, environmental and economic impacts. Ideally, sustainability is a balance which achieves positive outcomes (or at least minimises negative impacts) in all three areas, but social, environmental and economic outcomes cannot be measured in comparable terms. To compare them, arbitrary weightings must be used to equate fundamentally different quantities, and the choice of this weighting can skew the outcome in favour of one category. In a capitalist economy, environmental and social impacts are often undervalued (if they are valued at all), resulting in the most economically advantageous option being declared 'sustainable' to the detriment of the environment and society (Gibson, 2006; Hutchins & Sutherland, 2008).

The climate crisis, driven by increasing concentrations of atmospheric $CO_2$, is currently being addressed in most countries by setting goals of achieving **'net-zero'** ↗ $CO_2$ emissions by a specific date. In most sectors, including the construction industry, the approach is to make incremental reductions in $CO_2$ emissions to achieve net zero within a particular timescale. However, incremental reductions frequently result in very little net gain after economic and population growth, and **rebound effect** ↗, are taken into account. Carbon dioxide emissions provide a convenient, quantifiable measure of environmental impact: represented by a single number that is easy to measure, incrementally reduce, price and trade. The more complex notion of 'sustainability' has become conflated with $CO_2$ emissions. This ignores a wide range of other negative environmental and social impacts which are much harder to measure, but just as important to address.

If we take a step back from the incremental race to net-zero and critically examine current practice and emerging trends in the construction industry, do we see sustainable forms of construction emerging? Do we see materials and processes which could continue to be used for hundreds or thousands of years meeting future generations' needs just as well as ours have been met?

Current approaches to reducing the environmental impact of the construction and operation of the built environment are firmly focused on reducing $CO_2$ emissions whilst ensuring construction is rapid, low-cost and low risk. Reductions in operational energy requirements, for example by improving insulation levels and increasing airtightness in cold climates, are used to justify the use of environmentally damaging materials during construction. The overall $CO_2$ emissions for the 'life' of the building are low, but the wider environmental damage caused by extraction of raw materials and processing to manufacture steel and cement cannot be offset (Blanco et al., 2021).

'Modern methods of construction', a term used to describe off-site, digitally enabled prefabrication of building components, are offered as a panacea for the ills of the construction industry – with the potential to deliver high-quality buildings rapidly, with minimal waste, at a fixed cost. The negatives are rarely discussed: the environmental impacts of the materials used, the difficulties of future maintenance and modification of factory-built buildings, a reduction in employment in local building 'trades' in rural areas and small towns, and the risk of greater architectural homogeneity from the mass production of increasingly large building components.

Given the extensive environmental damage and deeply harmful social impacts that have occurred since the Industrial Revolution, is it enough to aim to gradually reduce $CO_2$ emissions to zero and otherwise carry on regardless, extracting finite resources from ever more fragile **ecosystems** ↗? Rather than doing 'less bad',

**Figure 1**
The OME: the Hub for Biotechnology in the Built Environment's experimental building

should we aim to do good, where 'good' is defined not from an anthropocentric perspective, but from the viewpoint of all living things? Could we fundamentally rethink our relationship with the natural world and work symbiotically with it? In doing so, could we start to restore natural ecosystems that are vital for humanity's long-term survival? Could we live in what McDonough & Braungart (2013) describe as a world of 'abundance and delight' rather than a world of limits?

This approach is known as **regenerative design** ↗ (Lyle, 1996; Blanco et al., 2021). Regenerative design requires a whole-systems approach to design, to develop systems with self-sustaining flows of materials and energy whilst addressing the needs of society and the natural world | **Figure 2** ↗ |. Regenerative design is so fundamentally different to mainstream, short-term, narrowly focused design approaches that we cannot get there incrementally. In the context of the built environment, regenerative buildings may not even be recognisable as buildings, and they are unlikely to be 'built' as we understand the word now – rather, they will be *grown*.

To have any chance of achieving this extraordinary shift in how we create and inhabit the 'built' environment, we need to understand the very significant barriers to doing this. These barriers span the entire development and implementation process from research culture and research infrastructure through to the risk-averse, highly regulated construction industry.

## Infrastructure for Collaborative, Speculative Research

↗ **Glossary**
embodied carbon
UK Building Regulations
biomaterial
biofabrication

New materials or construction technologies are usually developed at small scale in laboratories and workshops by materials scientists and engineers. They are developed for a specific application, usually replacing an existing building element with a new one with improved performance or reduced cost; for example creating a new insulation material which provides a greater level of insulation for a given thickness, or a new concrete mix with lower **embodied carbon** ↗. The performance requirements of the new material, such as strength, stiffness, insulating properties, water resistance and fire resistance, are clearly defined by the properties of the existing material or component, and by regulatory requirements such as the **UK Building Regulations** ↗ or similar documents in other countries. These requirements provide a detailed brief for materials scientists and engineers, but this process inevitably results in small incremental changes and does not provide the opportunity to fundamentally rethink how we construct and operate the buildings in which we live and work.

In contrast, the aim of the HBBE is to enable early-stage prototyping at architectural scale. Alongside small-scale material development and testing in laboratories and workshops, it is important to creatively explore the potential use of new materials and processes at large scale. This then enables designers and architects to interact with these new technologies in an architectural context and become involved in their development, allowing large-scale experimentation to feed back down into laboratory-scale development. For **biomaterials** ↗ and biological processes which are unfamiliar in the context of the built environment, this multi-scale exploration is vital as we lack the tacit knowledge required to presuppose how these biotechnologies can be incorporated into the built environment. Crucially, early-stage, large-scale experimentation provides the opportunity to reimagine how we meet the most fundamental physiological human needs – provision of shelter, warmth, food and water – whilst also exploring the potential for the built environment to meet higher needs such as self-respect, creativity, curiosity and aesthetic responses through people's interaction and engagement with novel materials and processes.

Whilst it may feel counterintuitive to start to fabricate at large scale before we fully understand what a material or system is capable of, we feel that this is the only way to move beyond incremental change and creatively explore how biotechnology can benefit the built environment. This is, of course, an iterative process which requires communication and understanding across scales and disciplines. To maximise the chance of success, a dialogue is required between large-scale prototyping and lab- or workshop-scale development such that the application concept

**Figure 2**
Could buildings function like complex biological ecosystems? Mixed woodland is an example of a dynamic, complex, diverse, multifunctional, self-sustaining ecosystem. In contrast modern housing is static, has limited functionality, is environmentally damaging during construction and operation, degrades over time, and is difficult to reuse at end of life.

allow great freedom for creative exploration, but it would not be a real-life test of the technologies that were developed. The OME was envisaged as a stepping stone to the next level of testing: integration of materials and systems within live building projects through collaboration with developers, architects, and clients.

The spatial concept for the building | **Figure 4 ↗** | was to create a small, self-contained domestic space (the 'studio apartment') on the first floor, consisting of a combined kitchen and living space and a shower room with toilet | **Figure 5 ↗** |. The studio apartment was primarily designed as a space for carrying out microbiome studies, exploring the role of ventilation, material surfaces and other interventions in the makeup of the microbial communities with which we cohabit (**Chapter 2.5**). In addition, the studio apartment provides a familiar space to engage with the public and explore their attitudes to biotechnology in a domestic context. The apartment is positioned directly above a microbiology laboratory, such that waste pipes from the shower, toilet and washing machine can deliver 'waste' to the laboratory for processing to generate energy and other materials (**Chapter 2.6**). We are not proposing that houses of the future will have a laboratory; these processes would be miniaturised and integrated within a house, as gas boilers and water tanks are currently, but for the experimental development of these processes a larger space is required.

**Chapter 2.5**
↗ Page 116

**Chapter 2.6**
↗ Page 130

Whilst the creation of the studio apartment was required for research studies which require a domestic environment, we also wanted to explore radically different ways of building. We therefore created prototyping and exhibition spaces in the OME which were designed to be as flexible as possible. Wall lining boards were mounted with visible fixings to allow them to be replaced with new biomaterials once they were developed | **Figure 6, top left ↗** |; services were exposed allowing straightforward modification (for example low-voltage, direct current wiring may be installed to make best use of electricity generated from urine in the lab, discussed further in Chapter 2.6); and double-height spaces and an exposed timber structure facilitates the fixing and suspension of prototypes | **Figure 7 ↗** |.

A mesh deck at first floor level links the studio apartment to the south-facing front façade of the building | **Figure 8 ↗** |. This space was originally conceived as a small social or exhibition space, but the polycarbonate façade and south-facing orientation makes the mesh deck warm and brightly lit. This makes it ideal for growing biomaterials and it was used to grow bacterial cellulose in bulk for the façade prototype (**Chapter 2.2**).

**Chapter 2.2**
↗ Page 76

Externally the OME is designed to become colonised with a range of prototypes exposed to external weathering on each of its four façades. The low-carbon precast concrete panels at ground floor level have robust, cast-in threaded fixings to allow prototype cladding and wall panels to be mounted on the façade | **Figure 6, top right ↗** |. Rainwater from the roof is very visibly collected in a single tank to enable living prototypes on and around the façade to be watered as required (**Chapters 2.3** and **2.6**). A section of wall on the east façade is timber clad and designed such that the entire wall can be removed, to allow complete experimental wall panels to be installed and tested in future.

**Chapter 2.3**
↗ Page 90
**Chapter 2.6**
↗ Page 130

**Figure 5**
Studio apartment – domestic spaces within the OME

Figure 6

Figure 7

**Figure 6**
OME details, clockwise from top left: removable cork and wood-wool internal lining boards; concrete panels with cast-in fixings and completely removable timber clad wall panel; south facing polycarbonate façade and prominent rainwater collection; interior view of the polycarbonate façade and exposed timber frame

**Figure 7**
Mesh deck area on the first floor of the OME

**Figure 8**
Flexible prototyping and exhibition spaces within the OME

Figure 8

## 2 Review of Objectives

Returning to the nine initial design objectives for the OME, it can be seen that two objectives were not achieved; 'integration of nature' and 'an exemplar in environmental sustainability' | **Figure 9** ↗ |. The OME provides a unique space for developing and testing new biotechnologies. However the building itself is constructed using conventional construction materials and methods: it makes limited use of natural or biobased materials, does not incorporate any living elements, and does not allow light and air to permeate the building as freely as originally envisaged for a building that aims to support growth. This occurred despite HBBE members being closely involved in the design process and advocating strongly throughout for the use of low environmental impact materials and less conventional forms of construction. Why was it so difficult to achieve our vision for this building?

| **OME primary design objectives** | **Comment** |
|---|---|
| An experimental laboratory to test biological systems and technologies | Objective achieved |
| A living zone to replicate habitable environments | Objective achieved |
| An exhibition and display facility | Objective achieved |
| A mobile OME unit | Objective revised – requirement for 'mobile OME' for public engagement events was later reimagined as a wider framework of participatory activities |
| Flexible and adaptable | Objective partially achieved – some flexibility, but many elements fixed |
| Integrate nature: living, growing organisms | Not achieved |
| A location to test interaction of concurrent systems and technologies | Objective achieved |
| Establish a clear and distinct identity for the OME | Objective achieved |
| An exemplar in environmental sustainability | Not achieved |

**Figure 9**
Review of OME design objectives

The close involvement of HBBE members in the design and construction of the OME highlighted multiple barriers to the adoption of non-standard materials and technologies, providing valuable lessons for the successful implementation of novel (bio)technologies in the built environment.

The original HBBE concept for the building was a timber-framed structure with a translucent ETFE roof, placed within the microclimate of a walled garden. The ETFE roof was chosen to flood the building with daylight to promote the growth of biological prototypes, whilst providing a 'semi-internal' prototyping space where temperatures would fluctuate diurnally. The boundary between inside and outside would be blurred, forcing us to reconsider whether we should be completely isolated from the outside world in terms of both light and temperature. At ground floor level a thick, solid brick wall was proposed to provide security and store thermal energy to passively moderate the internal temperature of the building. This wall extended beyond the building on the south façade to form a garden wall, creating a walled garden where external prototypes could be fabricated and displayed, and enclosing four existing trees. This concept was changed beyond recognition by a collision with the tenets of the construction industry: risk, regulation, time and cost.

The solid brick wall, timber frame and ETFE were all unfamiliar materials for the contractor, architect and engineers: this meant risk. For the construction industry, risk is minimised by using known suppliers, materials and construction processes. The result is great resistance to change. The perceived (or actual) risk of trying something new manifests as an increase in cost to mitigate the risk and ensure that a profit will be made. Trying to incorporate multiple unfamiliar materials resulted in inflated cost estimates, which meant that alternative materials were proposed for all these elements to keep the project within budget. Furthermore, the OME had to be built to strict deadlines to meet the requirements of the funder, and to ensure that the building was available for researchers to use. This meant a lack of time to explore and understand unfamiliar materials and construction processes, again driving the project towards known materials and supply chains.

A straightforward proposal by the HBBE to minimise the environmental impact of the building was to use timber whenever possible; in particular for the structural frame of the building. Months of negotiation were met with reluctance based on unfamiliarity. Careful use of terminology was vital – a 'timber frame' describes many different forms of construction, but for those unfamiliar with timber construction these distinctions could be lost, resulting in incorrect cost comparisons being made as non-equivalent items were compared. This demonstrated how easy it is for a contractor to manipulate the client into using familiar (low risk, high profit) materials by presenting alternatives in a negative light. Seeing this happen for timber, a well-established construction material, demonstrated just how challenging it will be to integrate fundamentally new materials and processes into building construction.

To comply with UK Building Regulations, the 'semi-internal' prototyping spaces had to be categorised as either 'internal' or 'external' so that the relevant requirements could be applied. If the space was 'internal' it had to meet all requirements for insulation, which the ETFE roof could not achieve. If the space was external there could be no form of heating within the space. Our intention was to have the space generally unheated, and use heating only on the coolest days, but this strategy did not comply with the Building Regulations.

The dense city centre location of the university meant that the building competed for space with other uses and was subject to strict planning rules. The proposed brick garden wall to the south of the OME was rejected by the City Council planning department as not being 'in keeping' with the surrounding landscape. On the south side of the OME site, there was a line of four trees only a couple of metres away from the proposed building. We were adamant that the first step in creating our experimental 'living' building should not be to cut down these trees. But they were cut down nevertheless – because there was a risk that they might be damaged during construction! If the trees had been kept, the contractor would have been legally required to ensure that they didn't damage them, including establishing a protection zone around them, which would have made construction very difficult. Instead they were permitted to cut down the trees and replace them with much smaller trees once construction was complete. Such regulations make it impossible to insert buildings into an existing living landscape.

## New Beginnings

Imagine if I invented a new construction material and I was presenting it to an audience of engineers, building contractors and architects at a major trade show. I start my presentation with images of a slender building frame made from a shiny, silver-grey material and proudly proclaim that this new material is fantastically strong and stiff in both tension and compression (unlike many materials which are weak in tension); it is ductile (it can bend and absorb energy; it won't crack and fail catastrophically if it is overloaded, making it very safe for the construction of buildings); and it can be recycled. The audience listen attentively; they are interested but cautious – they have heard plenty of similar claims over the years which have come to nothing. Hands are raised in the audience: 'That sounds like it has potential, but are there any problems with this new wonder-material?', 'Is it heavy?', 'Does it perform well in fire?', 'Is it durable?', 'Can we cut it to shape on site?', 'Can we source it locally?', 'What is the environmental impact of acquiring the raw materials?', 'Does it require a lot of energy to produce?'.

My answers leave a lot to be desired: it is extremely heavy, so the strength-to-weight ratio is no better than timber; in a fire the material rapidly loses its stiffness so it must be protected from fire with specialist paint, plasterboard or concrete; if the material is exposed to air and water it flakes and crumbles apart; it is difficult to cut and shape – components must be factory made and are very difficult to modify on site; the raw materials are only available in some countries so the materials must be transported globally; mining and processing the raw materials produces large quantities of toxic waste and uses vast amounts of energy (Conejo et al., 2020). The audience look disappointed but not surprised, and suggest that this new material isn't going to replace timber any time soon.

We are of course talking about steel, and a similar critique could easily be made for concrete. Yet these materials are used extensively and without question in almost every country in the world. They were the main constituents of the foundations, walls and structural frame of the OME. In fact, similar arguments can be made about every modern technology – from sewage processing to fast fashion, from cars to industrialised agriculture. It is increasingly well understood that they are predicated on an abundant supply of fossil fuels, excessive use of natural resources, and a disregard for humankind's impact on the natural environment. But what is less often discussed is their limitations. Returning to construction materials, new and traditional materials are routinely assessed against the best attributes of steel and concrete and are found wanting. Bamboo is not as strong as steel; mycelium composites are not as strong as concrete blocks. These comparisons are damaging and inappropriate for two reasons:

1. Alternative construction materials do not need to have the same properties as steel and concrete, because they do not need to be used to build in the same way. Each material should be used to build in the way that makes best use of its properties – this is the basis of vernacular architecture where form follows the capabilities of local materials, creating distinctive local architecture.

2. It is unusual for a comprehensive comparison to be made which considers the limitations of existing materials and the potential benefits of new materials and (bio)technologies. For example, whilst biomaterials may typically have lower strength and stiffness than conventional construction materials, they have a wide range of benefits including accessible, local, low-tech, low-cost production.

It is vital that we assess new materials and systems on their own terms, as part of a complete system rather than in isolation, and this comparison needs to consider a broad range of long-term environmental and social impacts and benefits.

The construction industry is constantly being challenged to innovate to become more efficient, to deliver buildings which generate less $CO_2$ (during construction and operation) more quickly and at lower cost. But fettered by a culture of small incremental changes, and a reliance on digital technology to deliver efficiency (for example through ever more advanced building modelling and off-site fabrication) there are very limited opportunities to fundamentally rethink how we build, operate, maintain and inhabit our buildings. The chapters that follow explore how incorporating biotechnologies in buildings could radically change our relationships with both the built and natural environments, and in doing so allow us to go beyond sustainability and net-zero and start to think and design regeneratively.

**Figure 10**
The OME – a space for prototyping and testing biotechnological prototypes

## Bibliography


Blanco, E., Pedersen Zari, M., Raskin, K., and Clergeau, P. (2021). 'Urban ecosystem-level biomimicry and regenerative design: Linking ecosystem functioning and urban built environments', *Sustainability*, 13(1), 404.

Brundtland, G. H. (1987). *Report of the World Commission on Environment and Development: Our Common Future*, United Nations.

Conejo, A. N., Birat, J-P., and Dutta, A. (2020). 'A review of the current environmental challenges of the steel industry and its value chain', *Journal of Environmental Management*, 259, 109782.

Gibson, R. B. (2006). 'Beyond the pillars: sustainability assessment as a framework for effective integration of social, economic and ecological considerations in significant decision-making', *Journal of Environmental Assessment Policy and Management*, 8(03), 259–280.

Hutchins, M. J., and Sutherland, J. W. (2008). 'An exploration of measures of social sustainability and their application to supply chain decisions', *Journal of Cleaner Production*, 16(15), 1688–1698.

Lyle, J. T. (1996). *Regenerative Design for Sustainable Development*. John Wiley & Sons.

McDonough, W. and Braungart, M. (2013) *The Upcycle: Beyond Sustainability – Designing for Abundance*. Macmillan.

Meadows, D. H., Meadows, D. L., Randers, J., Behrens III, W. W. (1972) *The Limits to Growth*. Potomac Associates – Universe Books.

# 1.2

Ruth Morrow
Louise Mackenzie

# 1.2

# Emerging Concepts in Biological Architecture

↗ **Glossary**
non-human
biofabrication
prototypes

This chapter is in search of a Biological Architecture. We are at the outset of a journey that will question long-held truths and practices in architecture, some of which have been questioned in broader theoretical and cultural critiques but which, so far, have failed to drive change in the built environment. One thing we do understand about Biological Architecture is that it puts a third party at its heart – a **non-human**↗, living entity – around which design practices, manufacturing and construction processes and our understanding of building use and maintenance will be required to evolve.

Biotechnology, biological engineering, synthetic biology and **biofabrication**↗ (there are many names emerging for the process of crafting with living material) are no longer the sole preserve of bioscientists with specialist skills. The technology has advanced sufficiently to enable mathematicians, engineers, computing scientists, architects, artists and DIY hobbyists to work with life as a mouldable material. This rapid transition from specialist knowledge to accessible technology has led to the development during the early twenty-first century of the 'bioeconomy', broadly defined as 'the production and use of biological resources, products, and processes to replace fossil resources and/or sustainably provide goods and services' and further, the 'circular bioeconomy', which conflates the former with a circular economic model where resources are reintroduced into the system (Kershaw et al., 2020). The substantial investment that is being poured into biotechnology, coupled with the pressure upon those working within biotechnology to find solutions to environmental challenges, risks meeting a perfect storm of hype and unrealistic promises. There is significant expectation for viable, scalable products to be brought to market in a field where the complexities of multi-species relationships are still largely untested beyond the laboratory.

Biotechnology, like all technologies, is part of the value system that drives and develops it and as such, can fall foul of disparity and privilege of access. The emerging concepts of bioeconomy and circular economy have led to government adoption of bioeconomic policies in many countries, yet these have tended to fall within existing infrastructural and institutional frameworks and thus have the potential to 'perpetuate respective socio-ecological conflicts and injustices on various scales' (Vogelpohl and Töller, 2021). Western-centric domination of dialogue, policy and practice around the emerging (circular) bioeconomy means a lack of voices from lower-income countries (Kershaw et al., 2020). The development of the bioeconomy on a global scale therefore is in danger of being shaped by existing capitalist models of production that are increasingly recognised as neoliberal extractivism (Backhouse et al., 2021).

Somewhere in between the environment of the biotechnology laboratory and the global infrastructure that will drive a bioeconomy, there is a requirement for space in which to build a link between the laboratory and the outside world. This space is both structural and relational. That is, physical spaces must be developed where biological materials and systems can be allowed to thrive in the real world, outside of the sanitised environment of the laboratory. But more than this, in order to approach the possibility of deriving new ways of working that do not fall into existing traps of production and extractivism, multiple voices must be invited to contribute to the making of biotechnological materials and practices.

In this chapter, we provide both the theoretical context and the relational approach within which the biotechnological **prototypes ↗** in this book sit. We set out five underlying concepts that these early tests suggest will be necessary for future biological architectures: Permanence to Impermanence; Shifting Scales; Embracing Moisture; Growing Aesthetics, Growing Values; and Responsible Interactions. We examine the ecologies of co-production that have emerged across the prototypes described in this book and explore the approaches employed by researchers to address the questions: How do we come together to build in a world of multiple perspectives? How do we generate and regenerate architectures that meet the needs of not just multiple peoples but multiple species? And what does it mean to develop architectures for planetary survival?

**Context**

## 1 Sites of Life

Writing for the 2020 exhibition 'Critical Zones' at the Center for Art and Media (ZKM) in Germany, philosopher and social theorist Bruno Latour described the mantle that forms the surface of the Earth and within which all life exists. 'Barely visible, it being only a few kilometres up and a few kilometres down at most. It is no more than a varnish, a thin mat, a film, a bio film. And yet, pending the discovery and contact with other worlds, it is the only site that living beings have ever experienced. It is the totality of our limited world' (Latour, 2020). Within our current climate emergency, this 'critical zone' has become an urgent site of examination and contextualisation as we begin to understand the damaging effects of unsustainable growth. Scientists, artists and academics across disciplines are exploring soil, air and water for insights that can help us understand our role in repairing and replenishing a damaged planet (Tsing et al., 2017). As architects of the built environment, we are part of the processes that extract from and build within this space. How, then, can we enhance our understanding of this critical zone before we intervene?

Chapter 1.1
↗ Page 18

Chapter 2.3
↗ Page 90

Architects are accustomed to reading and responding to the tangible elements of a site: its physical form, infrastructure and biodiversity. They also engage with the intangible, cultural aspects of a site and its historic and social aspects. This book speaks of elements and processes that are not visible and are thus less culturally accessible. Before the construction of the OME (**Chapter 1.1**), microbiologists swabbed and genetically screened the site in order to understand its existing microbial profile. As the prototypes described in this book were installed the swabbing continued, revealing the extent to which even harsh alkaline environments (concrete surfaces) are colonised by microorganisms (**Chapter 2.3**). These microsites of life in the built environment have only recently become known to us (Ross and Neufeld, 2015). We are at the start of this exploration of how best to respond architecturally to the pre- and co-existing life forms of an apparently empty site. If we are to take seriously Latour's description of the fragile skin upon which all life exists, we must proceed with caution.

## 2 Process, not Progress

Perhaps we can begin by reflecting that life is a constantly changing, time-based process. Our planet has evolved from, and is indeed still authored by, a vast plethora of microorganisms. This world, with its multiplicity of living beings, can also be recognised as a single living (planetary) 'being' that is continually shaped by many entangled lives. Acknowledging and considering multiple species, each with their own agency and time frames, allows us to reconsider humankind's relentless quest for progress – what cultural theorist Maria Puig de la Bellacasa describes as 'technoscientific futurity' or 'the persistence of a paradigm that associates the future with progress' (Puig de la Bellacasa, 2020). To illustrate the complexity of our progress-oriented focus, she cites the 'dust bowl' phenomenon of 1930s America, where intensive farming displaced a generation of Americans. Of the technological innovations that followed, whilst some enabled advances in soil conservation, others generated further soil exploitation. The drive forwards, frequently equated with economic growth, is also challenged by economist Kate Raworth, whose theory of doughnut economics questions whether progress and growth can prevail if we are to meet the needs of a growing population (Raworth, 2017). Bringing these wider contexts to the language of architecture enables consideration of a future in which there are no guarantees that progress, development and growth (as we currently understand these terms) will serve us, and those we share the planet with, well.

Can we then envisage an architectural future where the concept of growth becomes biological? Contemporary art practice can help us think anew when encountering the messy liveliness of biological material practices. Since the early 2000s, pioneering bio-artists have been bringing biotechnological materials out of the lab and into public spaces: arranging special licences to exhibit genetically modified artworks (Boland, 2013) and presenting and caring for engineered life in gallery spaces (Catts and Zurr, 2012). Contingent on working with biological material and bringing it into a public context is the fact that it has agency and as such, has the propensity to 'move, leak, smell, or worse, *react*' (Mackenzie, 2017). This presupposes the need for a shift in societal values if we are to embrace living biotechnological approaches to the built environment. Cultural theorist Marietta Radomska suggests a position on biological material that may help to pave the way for such a shift, which is to propose the acceptance of life as uncontainable (Radomska, 2016). If we can consider life as a series of flows, where some species give way to others in cycles of life, death and decomposition, then we can perhaps begin to accept biological materials as necessarily transitional and even transformative.

The realisation that biological materials both have agency and are transformative compels us to reconsider our approach to architecture at a fundamental level. Biology affords us the insight that all the materials that we make and work with are part of a larger, more complex system; where tending to the needs of living organisms, in life, death and decay, is an inextricable part of how we work and live with them. The acceptance of such complexity enables us to consider that the environment is not solely ours, nor has it ever been. As such, there is a clear need to fundamentally reassess our understanding of the processes by which we live, work and construct spaces, and even to question whether to move beyond the concept of 'building', to 'growing with'.

### 3 Growing the Definition of Technology

↗ **Glossary**
more-than-human

Responsible Interactions is one of the research themes within the Hub for Biotechnology in the Built Environment. At its core is a desire to evaluate the methods and practices of architecture as they begin to merge with the practices of biological science. The prototypes (MacroArchitectures and MicroAssemblies), set out in the following chapters, were conceived as a way to enable researchers from across science, architecture and design to collaborate and share ideas. In developing biotechnological prototypes, a wider framing of technology is required, starting from the work of Judy Wajcman, who developed a three-layered feminist definition (MacKenzie and Wajcman, 1985; Wajcman, 1991). The first layer is associated with the term 'technology'; that is, the object, hardware, software or the thing of it. The second layer is a form of knowledge that surrounds the 'thing' and arises during the processes of making, repairing and maintaining the thing. This is tacit knowledge, or 'know-how': a visual, tactile and contextual rather than a verbal, quantitative and abstract form of knowledge. The third layer is the interaction of people with know-how and the thing. For Wajcman, human interaction is an implicit component of technology. All three layers of the definition are interdependent. In order to examine and learn from the biotechnological prototypes included in this publication, we therefore expand Wajcman's definition to acknowledge that the first layer – that is, things – can also be living; that the second layer can include other tacit skills such as tending, caring and growing; and that the third layer also involves interactions with non-humans. Thus we move towards an *entangled* definition of a **more-than-human** ↗, ecological technology for Biological Architecture.

The next section discusses key concepts that have emerged through and around the making of the MacroArchitectures and MicroAssemblies described in this book, challenging some of the conventions of normative architecture and helping to inform our search for Biological Architecture.

## Emerging Concepts

### 1 Permanence to Impermanence

At the core of our cultural understanding of the 'art of architecture' is a profound sense of permanence, solidity and stability, captured in the Vitruvian term 'firmitas'. What we understand historically of architecture is invariably only that which remains standing. Architectural lineage, however, is as much within the compost below our feet as it is in the tomes in our libraries. Across the world's history and geography, built environments are both permanent and impermanent. Indigenous populations have long understood the transience of humanity in relation to the environment. This is evident in the living root bridges, grown and tended over decades by the Khasi people of Northern India, or by contrast the quick-to-erect-and-deconstruct Boma corral homes of the Kenyan Maasai (Watson, 2019). Indeed when we take into account informal settlements, refugee camps and sites of communal festivals, some of the largest constructed spaces are, at least at their outset, designed for impermanence. Yet the majority of architectural discourse remains aligned to cultural longevity, and building practices are focused on material performances that are built to last, with low, or preferably no, maintenance. This perception of architecture's immutability is thus not only false but costly, leading us to constrain and indeed shrink-wrap future cultures in today's norms.

Contrast this implied permanence and fixedness with the theoretical and philosophical underpinnings of biological material (that can be traced through Spinoza, Bergson and Deleuze) which prioritise the porous, open-ended and processual nature of life. Life, in its many constituent forms, arises, produces (to excess), decomposes and is continuously re-formed in multiple and uncontainable ways (Radomska, 2016). In the prototypes described later in the book we see technologies that draw from organic resources and can be returned to the Earth or fed into other ecological cycles. We learn of processes that encourage the local cultivation of materials with the potential to empower communities to sustainably generate their own energy. We are beginning to understand that the impacts of our material relationship with the world last far beyond the objects we create. It is through a return to the biological that we can nurture an understanding of the transient and mutable nature of future architectures.

## 2 Shifting Scales

### Scales of the Body

Historically, scale in architecture is said to relate to the scale of the body: '... the measures necessarily used in all buildings and other works, are derived from the members of the human body, as the digit, the palm, the foot, the cubit, ...' (Vitruvius, 2015). But biology reminds us that the body is more-than-human, in that the number of bacterial cells within the average person is roughly equivalent to the number of human cells (Sender et al., 2016). Furthermore, we – a collective bodily mass of human, bacterial, fungal, viral cells, and so on – are interconnected with other living systems: part of a local, and indeed global body (and therefore planetary biome). Hence the task for Biological Architecture is to have concern not just for the macro but also micro-ecologies, where the human body and its spatial enclosures are part of a continuum of living things.

The challenge then is to understand how these micro- and macro- processes and structures work at the scale of the built environment and what their impact will be. In the past 200 or so years, the world has faced an exponential population growth, resource depletion and waste production. Human-made materials are now so ingrained within the Earth's surface that they define our geological era: the Anthropocene (Stromberg, 2013). Our rapid rate of technological acceleration has brought with it environmental, social and economic issues of imbalance across the world. This is observed through the scale of land clearance for commodity production, causing an estimated 27 per cent loss of global forest between 2001 and 2015 (Curtis et al., 2018), where vast swathes of land given over to cattle farming and single-crop monocultures has led to, amongst other issues, soil degradation and a loss of biodiversity. As we move towards using more biological materials in the construction of environments, there is thus an imperative to remain hypercritical of both the means and scale of production and how these impact human and planetary health.

Notably, Vitruvius's first century BCE treatise, *De architectura*, did not remain at the scale of the body nor indeed the building. His concern also encompassed building materials; material production and resource location; mechanisms of energy production, defence and attack; mathematical systems; and stellar constellations. And so it is for Biological Architecture which must now further expand the scales of concern from the micro to the planetary.

### Scales of Operation

↗ **Glossary**
Scanning Electron Microscope (SEM)
X-Ray Diffraction (XRD)

The use of the phrase 'MacroArchitecture' in this book pays homage to the term 'macroorganism'; that is, 'an organism which may normally be distinguished with the unaided eye' (OED Online). Working with microbial material generally begins in the laboratory in tiny vials and dishes, where the nature of microorganisms in different circumstances are closely observed. A central aim of the research described in this book has been to move those materials beyond the lab and into the world. This has been integral to the development of MacroArchitectures, but it has not been without its challenges. Four of the MacroArchitectures described in this book – BioKnit, Bacterial Cellulose façade, Biocellular Concrete and Towards a Self-Sustaining Home – directly address the issue of production at scale by transitioning out of the laboratory and into public space. These prototypes meet the challenge of multi-species relationships in ways that extend beyond empirical evidence to embrace tacit knowledge of the world around us. Researchers learned to adapt their processes to environments that are not completely sterile. They found ways for materials to co-exist, even thrive, alongside what would be described in a laboratory environment as potential contaminants. The inevitability of microbial contamination is exemplified in space research where, for example, NASA clean rooms (sterile environments where spacecraft are built) have been found to harbour bacteria (Yong, 2018). Biological Architecture must accept these interrelationships and find ways to build *with life* rather than in spite of it.

Alongside a requirement to bring the microbial to the macro scale comes a need to work with technology at the nano scale. Engaging in these new scales of operation naturally requires the biological architect to acquire a more expansive range of tools and techniques in order to effectively capture and illustrate the multiscalar vision, processes, and outcomes of the work. The pages within this book tangibly demonstrate some of the shifts in scale that architecture will have to contend with to represent biological materials and data. **Scanning Electron Microscope (SEM)** ↗ images and **X-Ray Diffraction (XRD)** ↗ graphs illustrate micro-structures and the presence of chemical and biological compounds. Time-based images and diagrams capture growth and change. Ultimately, the 1:1 prototype is used not only to test the research at scale in a real-world context and over time, but crucially it also enables and stimulates lively interactions with the living environment as a whole, from experts and passers-by to insects and microbial extremophiles.

### Scales of Time

The impact that we have on the environment is bound up in time frames that expand beyond the human. The lifetime of a microorganism is short. *Escherichia coli*, the common laboratory workhorse, lives for around twenty minutes. This may lead us to believe that microorganisms are of little significance and that our use of them as a resource has limited or no consequences. However, the lifespan of microorganisms (plural) is limitless. Provided with sufficient nutrients, a microbial community can exist indefinitely, gradually evolving as a result of subtle environmental changes (Fox and Lenski, 2015). We can speculate that forms of life, from flora and fauna to bacteria and fungi, have no comprehension of a human-made world, other than to experience it as an environment within which to grow and adapt. In the following chapters, we learn of microbial ecologies that form on and within the spaces and materials that we create. A biological architect must therefore understand the ways in which the built environment inevitably leads to evolutionary change. Difficult though this may be, we also must take into account that such change will take place in more-than-human time frames that range from a few hours to many centuries.

Living with microbial ecologies and their evolution then also allows us to rethink the concept of maintenance in the built environment. Across all cultures, buildings have been, and in many instances continue to be, maintained seasonally: another layer of mud or lime render applied to walls, damaged shingles replaced, timber re-primed, walls whitewashed, and surfaces dusted. As our understanding of living materials moves to the molecular level, what we once considered part of building maintenance can, in some cases, now be understood as biological processes that we might learn to live with and tap into. One such example is the green slime found around leaking downpipes, which we now understand as an algal biofilm with a range of potential applications: from atmospheric $CO_2$ conversion to use as a fuel (Khan, Shin, and Kim, 2018). This kind of biological potential is explored in Part 2: MacroArchitectures where researchers recount tending to biological materials through the seasons, drawing our attention to the ways in which we might learn to embrace and indeed harness the seasonal changes of Biological Architecture.

## 3 Embracing Moisture

The phrase 'follow the water' encapsulates NASA's strategic search for life on extraterrestrial bodies and the degree to which water is considered fundamental to all life. Water is an effective solvent for chemical reactions and provides structural support to cells. It dilutes acid or alkaline environments that would otherwise be toxic; it is fundamental to the process of photosynthesis, reacting with carbon dioxide to produce glucose and oxygen; and it acts as a vehicle for both nutrient delivery and waste disposal. From space, the watery nature of the planet creates its signature blueness, while our bodies are composed of up to 60 per cent water. Our dependence on water as 'life's mater and matrix, mother and medium' (Szent-Györgyi, 1971) connects us not only to other people, lovers and children, but to other bodies of water and to non-humans in a 'more-than-human hydrocommons' (Neimanis, 2017).

The introduction of moisture to buildings creates something of a dilemma. In the built environment, water is generally channelled to and from buildings to service occupants' drinking, cooking and washing needs, and often to remove their waste. Otherwise, buildings are detailed and sealed to avoid water or moisture penetrating the 'damp-proof membrane'. The ingress and egress (due also to human activity) of water from buildings is a significant concern since it can trigger mechanical and chemical processes, as well as biodegradation by fungal organisms such as dry rot or black mould. Whilst we are naturally and constantly in the presence of microorganisms, such moulds have been linked to hypersensitivity in humans – for example, asthma, rhinitis and eczema (Vitte et al., 2022) – and even to serious illness and death (Brown and Booth, 2022). The resulting determination to keep water at a distance from human habitation, however, means that the process of constructing the built environment equates to desertification, turning once-living landscapes into ecologically extreme environments.

Biological Architecture seeks a nuanced approach that brings water purposefully back to the built environment. Working with biological material moves into the realms of what artist Roy Ascott has described as 'moist media' (Ascott, 2000). While Ascott used this term to describe the fusion of biology and computing science, it is equally applicable to the fusion of biology and architecture, where the moistness of a material relates directly to the liveliness of the organisms within it. Finding the balance between the necessary damp-free conditions for human and building health, and sufficiently moist environments for living materials is one of the core challenges of Biological Architecture (Alsmo and Alsmo, 2014). Whilst this research is still in its infancy, possible responses demonstrated by some of the prototypes in Part 2 and Part 3 show how we can allow for moisture regulation on and around the building façade through bioreceptive, biological and biotechnological materials. Such moist media might then become host to living micro and macrobiomes, enhancing the locale's biodiversity and counteracting the effect of urban heat islands through evaporative cooling cycles.

## 4 Growing Aesthetics, Growing Values

↗ **Glossary**
biomaterials

The expression of and social connection to biological architecture is in its earliest, most nascent stage. It will develop through continued interaction between organisms and humans, one where designers are no longer in control but are part of a process of collaborative emergence (Sawyer, 2003). When scaled up, the prototypes and **biomaterials** ↗ in this book created many technical and logistical challenges, and in parallel an entirely new set of sensory experiences were evoked. They were described using terms such as smelly, slimy, earthy and lumpy. The materials and systems oozed, sweated, burped and leaked, and their surfaces, once moist, became tacky, cracked and crazed. These are experiences that are not normally associated with the context of construction. Initially reactions fell into two categories: repulsion and curiosity. It was the latter that persisted and over time, curiosity was informed by an understanding of the potential environmental benefits. Gradually those who worked with the prototypes became accustomed and encultured to these new experiences, to the extent that the materials and processes became familiar, assuring and 'homely'. The work of Biological Architecture then is not only a process of transforming materials and processes, but also one of being transformed in our habits and attitudes towards them.

Anecdotally, we have found that exposing the ethical narratives around these materials triggers a more engaged and sympathetic aesthetic response in those who interact with them. People are as likely to respond to architecture as much through their ethical values as through their aesthetics. In Biological Architecture we aim to trigger both. Nevertheless we have to recognise that we are at a very early stage in the manipulation of these technologies. Their expression and resolution is currently crude and lumpy. If we consider that humankind has been working with timber for over 10,000 years it becomes clear that we have far to go in understanding, working with and refining our relationship to these new biotechnologies. At the same time, our growing understanding of how to create bioreceptive materials that encourage, support and sustain micro and macroorganisms makes us increasingly aware of the needless resources, effort and energy invested in making building materials that are smooth, clean, thin and forever 'young'. Despite, or perhaps in place of, our increasing fixation with cosmetic enhancements of our own bodies, can we learn to embrace architectures that are hairy, wobbly, pungent, and that grow ... old?

**Built to Last or Grown to Waste?**

This is not to say that newness does not have a place within Biological Architecture. The life cycle of biological materials presents an interesting possibility for how we adapt to live with biology in the context of the home. The BiomateriOME prototype (**Chapter 2.5**) categorises biological materials into The Living and The Lived. The latter we are all familiar with, comprising examples such as wood or cotton, but the former may comprise materials that are alive at the point of construction or even whilst in the built environment. While *The Lived* materials have a longevity and durability that we understand, *The Living* respond to our care and attention in new ways, and may in some cases deteriorate faster than we are used to. Thus we have the opportunity to think again about what we value in the way that we construct our environment and whether we can cherish objects that are not 'built to last' but rather 'grown to be de- (and re-) composed'. The concept of fast fashion may have a place in the world of biological material, if we can accept – and even value – the possibility of composting products as readily as we produce them. What would it mean to develop materials for the home that could be recycled every few years or even season by season?

**Chapter 2.5**
↗Page 116

## 5 Responsible Interactions

In both the MacroArchitecture and MicroAssemblies described in this book, the ways in which researchers discuss the living organisms they work with varies. For some researchers, they are simply materials; others describe facilitating biological processes, whilst others speak of being in collaboration. Improvisational theory provides some useful insights for managing these new and complex interactions between practitioners and living materials where no one organism is in control. 'Adaptive expertise' refers to knowledge and skills that are circumstantial and contextualised within a collaborative yet improvised event (Sawyer, 2003; Berliner, 1994). Adaptive-like, normative expertise is gained through practice and experience but is not asserted in collaboration. Rather, preparation and previous experience allows the actors to be *creatively responsive* in the context of the improvisation and, ultimately, adaptively expert with others (Morrow, 2021). In Biological Architecture, improvisation is a learning exercise between practitioner and living organism. In improvisation there is always an element of unlearning, in order to relearn in context, developing new layers of collective practice. Both practitioner and organism bring their 'expertise' to the improvisation, but what emerges comes from a process of adapting and accommodating, acknowledging the other and accepting that difference will result.

Across the projects a further theme emerges: the need for new language to describe these responsible interactions with biological materials and processes. Biological art practices have already navigated this terrain. Oron Catts and Ionat Zurr's Tissue Culture & Art Project, with its focus on human and animal material, raised the question of how to care for, and even how to kill, what they describe as 'the semi-living' (2003). Špela Petrič's aesthetic engagement with both animals and plants presents scenarios in which biological organisms are put to work, given space to play and even offered ways to improve their sex lives (Petrič, 2015, 2020, 2014). In the space of the microbial and molecular, the idea of care becomes more complex. Organisms are harder to relate to and yet we are in constant interaction with them.

In our search for a Biological Architecture, we learn that even the smallest living organisms tend not to do our bidding. We are slowly learning to move beyond the language of control to a language of reciprocity. Those who work with new biomaterials are close observers. They seek to understand both organisms they work with and the environments they thrive in, and are mindful of their by-products and waste streams (spores, acetic acid, and cellulose). Throughout this observation phase, researchers progress from being attentive to tending or cultivating organisms; that is, helping them to sustain life. Outside the lab, as we apply these 'biotechnologies' to prototypes, further responses are revealed – observing changes in context and across time, seeking to maintain and responsively assist the materials to adjust to the external environment. Living with the biological reinstates relationships of care in architecture, offering a greater connection between people and the resources that they live from. Just as we sense the biological through, for example, knowing when our food is out of date, or when to do the laundry from the smell of our clothes, we can become more attuned to and in rhythm with the biological home.

### Taking Uncertain Steps Together

The MacroArchitectures of this book represent the first steps in applying these new technologies, at scale, and in the world beyond the lab. They have allowed researchers to better understand both the opportunities and challenges ahead. As with all prototypes it is an iterative investigation involving trial, error and adjustments. Some problems that were initially anticipated failed to emerge whilst unforeseen challenges came to the fore. The next stage, already underway, is to return to the controlled conditions of the lab where smaller-scale, serial tests and trials are run and where bioengineered solutions are being developed by microbiologists.

As Biological Architecture develops it will naturally incorporate synthetic biology; that is, the manipulation of the genetic content of living organisms to enhance their contribution to a particular process or output. This genetic 'designing of life' might be termed radical evolution, where through human cultivation and propagation, new species are created at a pace never before witnessed, in what cultural philosopher Vilém Flusser has described as the Disneyland of the future (Flusser, 1988). One such example is the engineering of bacteria to produce microbially induced calcium carbonate precipitation – a process that some species of bacteria perform in nature, but which is hard to replicate in laboratory contexts. Researchers envisage using synthetic biology to create buildings which not only heal themselves, but do so in the tradition of Japanese wabi-sabi, referencing past trauma through cracks that microbes populate with contrasting colour schemes. Researchers are also working on engineering organisms to efficiently generate gas and electricity for domestic application. Looking further, we might conceive of walls that are biologically engineered to respond to heat and moisture, stretching to open vents or holes that reseal when the climate changes; surfaces that alert us to environmental hazards by giving off sharp odours; or façades that change colour in relation to pollution levels or the changing seasons. Further still, we can imagine buildings in any form, grown from materials engineered for suppleness, speed of growth, or even resistance to mould or decay.

Thus, crucially, Biological Architecture must interact responsibly with the ethical, political and ecological contexts that determine whether society is ready to accept the radical evolution that synthetic biology promises. The United States and other countries around the world already permit genetically modified organisms in food crops, and COVID vaccines created using genetic technologies have saved the lives of millions: therefore is it only a matter of time before genetically engineered life becomes part of the living environment? The concept of radical evolution must be read alongside the emerging concepts that we highlight in this chapter, as we consider how (or if) the new species that we are propagating on a micro scale in the laboratory will have macro consequences on a global stage.

In conclusion, we can only say that the journey ahead, like life itself, is complex, uncertain and spans evolutionary time. If the architecture of the future is to be biological, the next steps have to be ethically and critically considered, inclusive of all interactions and mindful of potential trajectories. Foremost, we need to acknowledge that we are not the only actors who shape the future of how we live.

## References


Alsmo, T., and Alsmo, C. (2014) 'Ventilation and Relative Humidity in Swedish Buildings', *Journal of Environmental Protection*, Vol. 5, No. 11.

Ascott, R. (2000) *Art, Technology, Consciousness: mind@large*. Intellect.

Backhouse, M., Lehmann, R., Lorenzen, K., Puder, J., Rodríguez, F. and Tittor, A., 2021. Contextualizing the bioeconomy in an unequal world: biomass sourcing and global socio-ecological inequalities. In *Bioeconomy and Global Inequalities: Socio-Ecological Perspectives on Biomass Sourcing and Production* (pp. 3-22). Cham: Springer International Publishing.

Berliner, D.C., 1994. Expertise: The wonders of exemplary performance. *Creating powerful thinking in teachers and students*, pp.161-186.

Boland, H. (2013) *Art from Synthetic Biology*. Thesis, University of Westminster.

Brown, Mark, and Robert Booth. 2022. "Death of Two-Year-Old from Mould in Flat a 'Defining Moment', Says Coroner | Housing | The Guardian." *The Guardian*, 2022. https://amp.theguardian.com/uk-news/2022/nov/15/death-of-two-year-old-awaab-ishak-chronic-mould-in-flat-a-defining-moment-says-coroner. (Accessed: 15 November, 2022)

Catts, O., and Zurr, I. (2003) 'The Ethical Claims of Bio-Art: Killing The Other Or Self-Cannibalism', *Australian and New Zealand Journal of Art* 5 (1): 1–19.

Catts, O., and Zurr, I. (2012) *Crude Life – The Tissue Culture & Art Project* (Retrospective catalogue), Lazina Centre for Contemporary Arts, Gdansk, Poland.

Curtis, P. G., Slay, C. M., Harris, N. L., Tyukavina, A., Hansen, M. C. (2018). 'Classifying drivers of global forest loss', Science 361, 1108–1111.

Flusser, Villem. 1988. "Curie's Children." *Art Forum* October.

Fox, J. W., and Lenski, R. E. (2015) 'From Here to Eternity – The Theory and Practice of a Really Long Experiment', *PLOS Biology* 13(6): e1002185.

Kershaw, E.H., Hartley, S., McLeod, C. and Polson, P., 2021. The sustainable path to a circular bioeconomy. *Trends in Biotechnology*, 39(6), pp.542-545

Khan, M. I., Shin, J. H., and Kim, J. D. (2018) 'The Promising Future of Microalgae: Current Status, Challenges, and Optimization of a Sustainable and Renewable Industry for Biofuels, Feed, and Other Products', *Microbial Cell Factories* 2018 17 (36).

Latour, B., (2020) Bruno Latour on CRITICAL ZONES. ZKM. 2020 https://zkm.de/en/zkm.de/en/ausstellung/2020/05/critical-zones/bruno-latour-on-critical-zones (Accessed: 31 October 2022).

Lea, T., (2015) 'What has water got to do with it? Indigenous public housing and Australian settler-colonial relations', *Settler Colonial Studies* 5(4), 375–386.

Mackenzie, L. (2017) *Evolution of the Subject: Synthetic Biology in Fine Art Practice*. Doctoral Thesis, Northumbria University.

MacKenzie, D., and Wajcman, J. (eds.) (1985) *The Social Shaping of Technology. How the refrigerator got its hum*. Milton Keynes; Philadelphia: Open University Press.

Morrow, R., and Waddell, T. (2021) 'Engaged Practices: learning from improvisation', in Ferdous, F., and Bell, B., *All-Inclusive Engagement in Architecture. Towards the Future of Social Change*. Routledge, New York.

NASA, Mars Exploration Program and Missions Overview. https://mars.nasa.gov/programmissions/overview/ (Accessed: 31 October 2022).

Neimanis, A. (2017) *Bodies of Water: Posthuman Feminist Phenomenology*. London: Bloomsbury Publishing.

OED Online, Oxford University Press, September 2022, www.oed.com (Accessed: 19 November 2022).

Puig de la Bellacasa, M. (2020) 'Soil Times: The Pace of Ecological Care', in *More-Than-Human*. Het Nieuwe Instituut, Office for Political Innovation, General Ecology Project at Serpentine Galleries and Manifesta Foundation.

Radomska, M. (2016) *Uncontainable Life: A Biophilosophy of Bioart*. PhD Thesis, University of Linkoping.

Raworth, K. (2017) *Doughnut economics: seven ways to think like a 21st-century economist*. London: Random House.

Ross, A. A., and Neufeld, J. D. (2015) 'Microbial biogeography of a university campus', *Microbiome* 3, 66.
Sawyer, R.K., 2003. Emergence in creativity and development. *Creativity and development*, pp.12-60.

Sender, R., Fuchs, S., and Milo, R. (2016) 'Revised Estimates for the Number of Human and Bacteria Cells in the Body', *PLOS Biology* 14(8).

Špela Petrič." n.d. Accessed November 17, 2022. https://www.spelapetric.org. (Accessed: 17 November 17, 2022).

Stromberg, J. (2013) 'What Is the Anthropocene and Are We in It?', *Smithsonian Magazine*, 2013.

Szent-Györgyi, A. (1971) 'Biology and Pathology of Water', *Perspectives in Biology and Medicine* 14(2), 239–249.

Tsing, A. L., Swanson, H. A., Gan, E., Bubandt, N. (eds.) (2017) *Arts of Living on a Damaged Planet: Ghosts of the Anthropocene*. Minneapolis: University of Minnesota Press.

Vitruvius Pollio, M. (2015) 'The Architecture of Vitruvius, Book III', Gwilt, J. (Trans.), T*he Architecture of Marcus Vitruvius Pollio: In Ten Books*. Cambridge: Cambridge University Press.

Vitte, J., Michel, M., Malinovschi, A., et al. (2022) '*Fungal exposome, human health, and unmet needs: A 2022 update with special focus on allergy*', Allergy 2022, 77: 3199–3216.

Vogelpohl, T. and Töller, A.E., 2021. Perspectives on the bioeconomy as an emerging policy field. *Journal of Environmental Policy & Planning*, 23(2), pp.143-151.

Wajcman, J. (1991) *Feminism confronts technology*. Cambridge: Polity.

Watson, J. (2019) Lo–TEK, *Design by Radical Indigenism*. Taschen.

Yong, Ed. 2018. "Bacteria Survive in NASA's Clean Rooms by Eating Cleaning Products." *The Atlantic*, 2018. https://www.theatlantic.com/science/archive/2018/06/bacteria-can-eat-the-cleaning-products-nasa-uses-to-sterilize-its-spaceships/562016/. (Accessed: 24 November, 2022)

# 1.3

Peg Rawes

# 1.3

# Kind Matters
## Ethical Approaches to Architectural Research

↗ **Glossary**
more-than-human
non-human
biodiversity

This chapter examines how a relational concept of matter supports ethical approaches to interdisciplinary architectural research, especially design that addresses fragile social, environmental and climate-changed inhabitation. Drawing from feminist and intersectional[1] practices, I highlight the benefits of these material modes of thinking for artistic, environmental, biological and engineered concepts of design. In addition, given the climate crisis and increased global levels of societal inequality, I suggest these approaches constitute valuable 'planetary relations' through which ethical individual and collective designs can be created that address today's urgent human-made environmental threats to life. This chapter does not focus on single or autonomous 1:1 built outcomes of architectural research. Rather, it suggests that relational practices emphasise the intersection of social, cultural and environmental ethics in design, at all scales and durations.

1
'Intersectionality' is a critical race theory term, coined in 1989 by black American legal theorist Kimberlé Williams Crenshaw, to address the combined lived experience of racial, gendered, sexual and age-related discriminations. Much recent critical race theory also extends to environmental and economic discriminations.

My discussion with the editors for this volume has taken place over the twelve months during which the UK hosted COP26 in November 2021. It is therefore written alongside the most recent UN-led effort to improve the 'design' of governmental, financial and corporate responses to the climate crisis. However, as was much reported during and since the summit, its success in achieving the necessary financial and political commitment to reduce carbon emissions has continued to concern those scientific and environmental communities engaged in tackling climate change. This failure to implement strong carbon emission restrictions now – rather than in ten, twenty or thirty years' time – means the catastrophic impacts of human-made climate change on vulnerable regions and peoples most affected by rising global temperatures is inevitably increasing. And, as we have recently seen in the UK, Europe and the US, climate change is already affecting the global north; for example, the extreme heatwaves in Spain, France and in the UK. Yet, here in the UK, and in the US, strong political action to address climate change and engagement with scientific evidence is missing. For example, the UK Government's plans for net-zero have been judged unlawful (Smith, 2022), and the most recent UK Government briefing on climate change was attended by a mere 10 per cent of UK MPs (Horton, 2022). In the US, the Supreme Court recently dismissed scientific climate change evidence in favour of protecting its dependency on the domestic fossil fuel industry (Milman, 2022).

However, as feminist and environmental research since the 1960s has amply shown, these are planetary matters which also highlight the uneven relational distribution of economic, social and environmental powers across the globe. They are, in current terminology, evidence of the '**more-than-human**' ↗ composition of our planet; of the fragile and damaged relations between human-built and **non-human** ↗ natural environments; and the exclusion of, or discrimination against, vulnerable peoples by political economic systems which deem them expendable or 'non-human'. For those who work to protect disadvantaged communities and remove structural inequalities, these are matters of deep ethical concern. Below, I consider the historical and current work of researchers and communities whose ethical practice benefits social and environmental matters, as a whole. Hence, my focus on 'kind matters' and 'matters of kind' draws attention to structural formations of living architectural habitats, and the need for architecture and the built environment sector to actively commit to, and undertake, ethical forms of design.

The phrase 'kind matters' speaks of ethical practices which protect and nurture equitable social and environmental relations. Such practices require the designer to 'care for' or 'take a care' in developing designs through which social and environmental equities are redistributed to those who are currently, and who have historically, been excluded from environmental security. The phrase therefore also intentionally underscores that kindness towards those most vulnerable to environmental and societal risk *matters* in design. The second term, 'matters of kind', takes account of cultural and physical *material* differences which structure organic and inorganic entities. It speaks to the importance of planetary **biodiversity** ↗, and of inclusive – that is, multi-racial, gendered, sexual, abled and aged – humane societies. In each phrase, the word 'matter' also captures the asymmetrical intersections that form historical, political and economic power relations (e.g. between the global north and global south communities), and the political economic organisation of these social relations.

Rightly, the architectural and built environment sectors are currently under substantial critical scrutiny for their role as professions and disciplines that construct the relationship between lived experience and the fabrication of our environment. Individual, community, socio-economic, political and health inequalities present in the built environment are therefore also, in part, the responsibility of the sector. At present, given the increased complex geopolitical uncertainty which is affecting different global communities – including devastating fires and floods; a third year of the COVID-19 pandemic; the Russian invasion of Ukraine; and the resulting energy and grain supply shortages – we might very well agree that our 'house is on fire'.

This imagery was given to me by a colleague, Camillo Boano, in the week following the Grenfell Tower fire in June 2017. In his book *The Ethics of a Potential Urbanism* (2017), Boano examines the neoliberal formation of advanced urbanism. He draws from the Italian philosopher Giorgio Agamben's political writings on art, architecture and society, including Agamben's image of a house on fire: 'it is only in the burning house that the fundamental architectural problem becomes visible for the first time. Art, at the furthest point of its destiny, makes visible its original project' (Giorgio Agamben, *The Man Without Content* [1999], cited in Boano, 2017, p. 25).

Agamben's words vibrated and blurred as I read them alongside watching the distressing images and reporting from the tragic fire and loss of life in Grenfell Tower, a local authority housing block in Kensington in the wealthiest borough in West London. The much-repeated phrase at the time – 'duty of care' – still jars today, given the systematic decline in British housing construction and welfare over the past fifty years. Rather than describing strong ethical agreement and action in the governmental and corporate design of society, the Grenfell disaster more accurately highlighted the British social housing sector's lack of accountability in tackling poor design, fabrication and construction standards. Five years later, at the close of hearings in the Grenfell Tower Inquiry, witness and expert testimonies have revealed the horrific loss of home and family, and neglect of residential life by regulatory, governmental, architectural, engineering and building construction industries (Grenfell Tower Inquiry, 2022).

## Biopolitical Relations

Grenfell provides an example of contemporary neoliberal and corporate failings in ethical responsibility for the built environment. Unsurprisingly, the architectural profession is also being criticised for its weak ethical regulatory and employment practices, which reproduce similarly damaging 'matters of kind'. In contrast, evolutionary biologist/feminist philosopher Donna Haraway's term 'response-ability' offers a 'situated' concept of agency which acknowledges the unequal power relations of global capitalism (Haraway, 2016, p. 35). In recent years, multigenerational environmental architectural communities have also called for urgent structural changes to the profession's employment and educational cultures for example, activist and lobbying groups such as The Architecture Lobby, the Section of Architectural Workers (SAW) and Architects! Climate Action Network (ACAN). These have joined longstanding resistance by feminist, anti-racist and LGBT+ practitioners inside and outside the discipline to the structural reproduction of inequality. Collaborative groups, such as Parlour (Australia), and previously, Matrix Architects (UK), have promoted ethical pedagogies, processes and designs which address the damaging social consequences of discrimination. Today, their criticisms are reflected in planetary inequalities, including the COVID-19 pandemic, employment and housing precarity, and conflict-based and climate-changed vulnerabilities.

Again, by emphasising an ethical approach to 'more-than-human' relations, Haraway's term 'kinship' (Haraway, 2016, p. 3) provides a concept of social creation which puts 'matter' back *in companionship with others*. Judith Butler's etymological retrieval of the term 'matrix' for the word 'matter' links physical matter to maternal ethics, through which the individual and societal relations are affirmed (Butler, 1993, p. 31). In each word, these thinkers emphasise the creation of societies in which positive and inclusive attributes of human difference are *put back into* matter. Both writers argue that, by improving the power of these material ethics, more humane, representative and equitable forms of society are achievable, which benefit society as a whole. The importance of ethical collaborations for the formation of ethical societies is prioritised. In addition, these are values which matter especially for communities whose lives are historically and politically diminished or excluded by normative or repressive (e.g. neoliberal) social relations.

These matters of kind are also *intentionally* artful, celebrating the poetic sensibility of matter ('sciences of the sensibility' [Rawes, 2008, p. 84]). Such creative intersections therefore provide understandings that the world and its matters – buildings, bodies and ecologies – constitute lived corporeal relations, rather than objects derived from disembodied scientific reasoning. As such, idealised design intelligence which prioritises short-term normative or self-serving outcomes is replaced by thinking-as-living: *always* embodied; located in specific times, histories, spaces and places; and undertaken by specific communities and individuals. Consequently, architectural thinking shifts design to a *response-ability* for constructing life-worlds that protect the planetary relations underpinning human-made and non-human habitats and life.

Like Haraway's witty and transformative 'companion species', which celebrates the power of living in 'kin-ness' with the non-human (Haraway, 2008), ecologist Rachel Carson's ground-breaking exposure of the human damage to non-human environments in *Silent Spring* (1962) is still exemplary of this ethical design intelligence. But ecological philosopher Lorraine Code reminds us that Carson was nevertheless rejected by science for her 'vulnerability'; her sex and early death from cancer resonating with today's discrimination of those considered to be 'other' or medically vulnerable (Code, 2013, p. 83–4). Both Carson and Haraway's artful, intersectional science show the power of humane critique within and for our contaminated histories; for example, Haraway's exposure of the inhuman within the emancipatory modern technoscientific revolution, which *was, and still is*, responsible for diminishing and colonising the rights and lives of women, people of colour and disadvantaged communities (Haraway, 1997, p. 3). Again, given the resource depletion and extraction, and the social and environmental injustices which result from global neoliberal political economies, more effective ethical practices in architecture and the built environment sectors therefore remain essential.

## Stories From Science

Other artful economic, biological and geological writing follows Carson and Haraway's 'social science' example: some mythical, some scientific, some historical. For example, Donella Meadows et al.'s *The Limits to Growth: A Report for the Club of Rome's Project on the Predicament of Mankind* (1972) is celebrated for its macroeconomic analysis of world capitalism. But the book's post-war warning against unrestricted growth is also a critique of the danger accelerating fossil fuel markets bring to the security of modern political economies. As such, the book represents a preview of our failure to deal with fossil fuel dependency, and of its concomitant damage to human, environmental and planetary wellbeing.

More recently, environmental journalist Elizabeth Kolbert's *The Sixth Extinction: An Unnatural History* (2014) is indebted to Carson's storytelling in its narratives about the terminal loss of biodiversity. Geographer and geologist Simon Lewis and Mark Maslin's *The Human Planet: How We Created the Anthropocene* (2018) tells a history of the Anthropocene, underlining how scientific, economic and political extractions of environmental value and matter have directly impacted life, in all its forms, since the early humans. Following other Anthropocene researchers, they show how early modern agricultural capitalism captured land, peoples and grain, inaugurating one the most devastating forms of human-made climate change. Critical race theorist Katherine McKittrick's *Dear Science and Other Stories* (2021) articulates more recent linguistic, social and corporeal heritages of racialised capitalism. Reading 'against the grain', her poetic and critical analysis of the racism in modern science shows how a new humane agency can be created out of the inhuman enslavement of peoples. Thus, rather than excluding science from political discussions of lived experience, these writers' stories (including those who explicitly define themselves as natural or physical scientists[2]) emphasise the need to address both ethical and unethical scientific relations in protecting and nurturing equitable human and non-human matters.

While Anglo-American research cultures use the term 'discipline' – rather than the word 'science', as used in European academic traditions – history, despite claims made for the objectivity of historical facts, is not a neutral science. Indeed, as the researchers above show, it never has been. Rather, history is always, inherently, tied to the political imagination. For those writing Anthropocene histories about

**2**
Lewis and Maslin are examples of scientists who also use their public platforms to support citizen and activist critiques of climate change.

the neoliberal, human-changed formation of planetary life – and, I would argue, by extension, histories of architectural design – we therefore have a choice about how our histories address racial, gendered, sexual or economic inequalities within our societies. An earlier historical example of this agency is Baruch Spinoza, the seventeenth-century philosopher of life, who argued scientific reasoning is not confined to our intellectual powers. Rather, in *Ethics* (1677) he describes how our powers of reasoning are intimately connected to corporeal experience. For Spinoza, rational thinking is therefore directly related to our affective capacities for producing ethical individuals and social agency (i.e. our response-ability).

When taken as a canonical discipline, architecture has not endorsed the experiences of communities whose lives are seen as minor to the pursuit of modernist success. But, by listening to histories in which matters of kind *matter*, we can be changed by them: for example, Paul Gilroy's magnificent anti-racist thinking which, alongside other critical race practitioners, speaks of the 'roots' and 'routes', and the 'special stress that flows with the effort involved in trying to face (at least) two ways at once' (Gilroy, 1993, p. 3). Gilroy's concept of the *Black Atlantic* 'addresses ... the stereophonic, bilingual, or bifocal cultural forms originated by – but no longer the exclusive property of, blacks dispersed within the structures of feeling, producing, communicating, and remembering'. Such a history also requires architecture to practice equity and inclusion, to repair colonial exclusions, and to remove the racist structures from its educational and professional cultures. By doing so, black, coloured and indigenous communities who have experienced racism and discrimination may be enabled to be 'perceived as agents ... with an intellectual history – attributes denied by modern racism' (Gilroy, 1993, p. 3). These 'matters' are also present in the paintings of Frank Bowling: for example, in *Middle Passage* (1970), which remembers his own, and others', Windrush migration from Guyana and the West Indies to post-war Britain. Today, for many in these communities, the ongoing UK Government's discrimination continues their painful memories. As such, Bowling's paintings check the idealisation of notions such as 'home' or 'dwelling' in built environment discourses and education, which ignore the complex material histories of British society's post-war colonial and immigrant heritage.

## COVID Matters

Thinking about matters of life during the COVID-19 pandemic has also raised important questions about vulnerability and ethics in architectural practices. While writing an article on visual vocabularies in art and science in April 2020, after having caught the virus a few weeks earlier, the topic took on new and unsettling psychic, physical and planetary resonances (Rawes, 2021).

Over the past two years, the pandemic has brought into even greater focus differences in societal equity, especially for disadvantaged communities of colour, poverty, disability and receiving care. Life has been lived differently across worldwide societies and communities. The pandemic has shown that those with financial and political advantage have experienced some restrictions to their liveability. But for those who already experienced structural social inequity, the pandemic has further increased these issues, including declines in access to secure education, employment or health services. The pandemic has starkly shown us the extent to which relationships between self and society have increased proximity to, and experiences of, vulnerability. These social planetary inequalities can be characterised as 'crisis-ordinariness', a term coined by critical theorist Lauren Berlant to describe the precarity of our societal infrastructures (Berlant, 2011, p. 10). Heightened differences in liveability, and in communities' social and political agency, therefore constitute a biopolitical form of modern society. In turn, as a social discipline, this places an additional new response-ability on architectural design's creation of liveable social and corporeal relations at home and work, in our cities and towns, and in our institutions and cultures.

Feminist philosophers draw attention to the benefits of empathic justice for increasing social wellbeing. We might call this a kindness for humane matters. Judith Butler defines this kindness as the power of speaking for vulnerability. She describes it as 'a relation to a field of objects, forces, and passions that impinge or affect us in some way'. She continues, arguing that we need to:

> rethink the relationship between the human body and infrastructure ... to call into question the body as discrete, singular, and self-sufficient, and ... understand embodiment as both performative and relational, where relationality includes dependency on infrastructural conditions and legacies of discourse and institutional power that precede and condition our existence (Butler, 2016, p. 25).

In a talk given during the year before she died, titled 'How to Think About Death: Living with Dying' (2012), British philosopher Gillian Howie also examined this power of critical thinking, specifically for those living with life-limiting illnesses. She argued that such life-stories demystify discrimination of the vulnerable and, simultaneously, affirm the singularity and collective agency of individuals and communities who live such lives.

While we are still counting the cost of the initial and subsequent waves of the pandemic, it has also led to rapid redesigns of soft and hard architectures of life, such as new hospitals and vaccine technologies. Existing drugs have also been used to treat patients requiring mechanical ventilation, including Dexamethasone, a cheap cortico-steroid with anti-inflammatory properties. An immune suppressant, 'Dex' is also used to treat blood cancers, reducing inflammation and chemotherapy sickness. But, taken in high doses or for extended periods of time, its side effects include fierce mood swings, increased appetite, weight gain, insomnia, nausea and high blood pressure. Such matters of kind also draw from my own relational architectural practice of 'living-with' Dex as part of my caregiving over the past ten years, and in assisting the life and work of the artist I live with.

Tom Corby's art practice examines the intersections between bodies, environments, climates and their data. Produced during extended periods of self-isolation during repeated oncology treatments (akin to the UK's 'shielding' during lockdowns), his project, *Blood and Bones: Metastasising Culture* (2013–2016), visualises his daily affective modulations of 'living with' cancer, including his response to combination drug treatments and Dexamethasone in preparation for a stem-cell treatment. Drawings, graphs, photography and blogs chart his psychological and physiological 'data', the rise and fall of his blood platelets and his immune system response to the therapies (Corby, 2016). These art-full expressions of living-with disease resonate afresh in our new architectures of vulnerability and care of the pandemic. After experiencing the social and spatial architectures of lockdown – the daily governmental briefings streamed into our homes – returning to look at Corby's artworks now provides new insights of the dis-ease experienced in living with life-limiting illnesses.

From the macro-scale of a governmental management of the pandemic to the micro-experience of daily medical treatment, Corby's images and UK COVID-19 data represent two scales of biopolitical life. Both show the personal impact of self-isolation for those classed as 'vulnerable' during the COVID-19 lockdowns and during cancer treatment. On a macro-societal scale, the pandemic and oncology care generate data – information which now signals the potential financial value that can be extracted from personal data by financialised international healthcare provision. Also, within the UK Government's responses to the pandemic, its inconsistent use of the phrase 'follow the science' has highlighted how scientific evidence and expertise can be manipulated for short-term political gain. Social uncertainty and structural inequalities between different cultural, economic and regional communities have been amplified (e.g. the higher risk of infection and deaths for ethnic communities employed in the NHS, healthcare and public services [UK Gov, 2021]). As such, in the UK, independent scientific knowledge and the national health service provision have been tied together to form an intensely biopolitical value system.

As I mentioned above, writing about the politics of data visualisation during the first months of the UK lockdown reshaped my understanding of uncertainty in living with a vulnerable person. Combined, we might say that Corby's artworks and visualisations of British COVID-19 infections and deaths represent *bodies of work* about life-limiting illnesses. Each carries in it poignant biological and political embodiments of vulnerability and uncertainty. Gillian Howie's writing advocates learning from those who have life-limiting illness. Illness may mean mental and physical time and space are fearful and deeply isolating. But there are also times and spaces when acknowledging *dis-ease* can lead to agency for oneself, work, family, community, friendship, politics and poetics. Time is lived differently: not having time means that the individual's powers of choice and agency are intensified. Sometimes, for some, dis-ease can be put to work, and can make work for oneself and for others.

For architectural practitioners, these stories of 'living-with' and the respective 'bodies of work' made during such lives, may enable us to understand how life is designed simultaneously at multiple scales of agency – from addressing climate crisis and the pandemic, to giving and receiving care. Such understandings of lived experience can also inform historical or scientific knowledges in architecture: for example, practices of vulnerability may bring out important – but perhaps not-yet articulated or valued – poetic and ethical ways of knowing. In turn, understanding that architecture is formed out of ontological relations, including ones that are fragile or need repair, may stimulate us to better address the *dis-eases* of social and environmental inequalities, and to create more ethical practices of architectural care in which diverse intergenerational communities can flourish.

Finally, moving out to consider other multiple planetary scales, environmental artist Agnes Denes's 'Book of Dust' (1972–88) examines a constellation of matters of kind. Bodily, pharmacological, organic, toxic, climatic and planetary matters embody the agency of dust; for example, in the section titled, 'Nuclear dust', she writes:

> A nuclear war creates dust clouds caused by H-bomb blasts and by burning cities, forests, and industrial plants. Even a small war would produce a temperature drop sufficient to disrupt the world's agriculture. A massive war would create a frigid long nuclear winter.
>
> Amount of dust in air: up to one cubic mile for a 10,000-megaton war
> Composition: dust containing soot (carbon) and pulverised soil
> Temperature drop: below freezing in summer (-20 °C)
> Decrease in sunlight; sun appears thousands of times dimmer
> Deaths: people who survive initial blast die from prolonged starvation (Denes, 2008, p. 42).

Denes's aphorisms capture the power of material relations. Vibrating with material difference – from the crystal, nuclear, and biological to the elemental – her ecological matters compose human and non-human differences. Powers of matter, or matters of kind, structure planetary life-worlds. Matters, such as carbon emissions since 1870, climate-changed fires and floods, metabolic rifts and soil disruptions compose transformative 'bodies of knowledge' which have been, and will continue to be, evidence of human damage to environments. Since Denes wrote these words three decades ago, we have entered the time of irreversible tipping points. But these are also matters of kind in which the world and its matters – oceans, bodies and ecologies – suggest otherwise. Following Donna Haraway's 'response-ability', these are ethical matters of kin and of kindness, acting simultaneously across social, environmental, planetary and climatic scales. She, together with other thinkers and storytellers of the soil – the Chthulucene (Haraway, 2016, p. 2) – articulate non-human relations through vocabularies which address the damaged realities of our neoliberal and contaminated inequalities, but also speak of support structures which are available for enabling more ethical ways of living humanely with human and non-human worlds.

**Figure 1**
The texture and translucence of bacterial cellulose installed on the BioKnit prototype

## References


Baruch de Spinoza, 1677. *Ethics*

Berlant, L. (2011) *Cruel Optimism.* Durham CA.: Duke University Press.

Boano, C. (2017) *The Ethics of a Potential Urbanism: Critical Encounters between Giorgio Agamben and Architecture*. London: Routledge.

Butler, J. (2016) *'Rethinking vulnerability and resistance'*, in Butler, J., Gambetti, Z. and Sabsay, L. (eds.) *Vulnerability in Resistance.* Durham CA.: Duke University Press.

Butler, J. (1993) *Bodies That Matter: On the Discursive Limits of Sex.* New York: Routledge.

Carson, R. (1999) *Silent Spring.* Harmondsworth: Penguin Books.

Code, L. (2013) 'Manufactured uncertainty': epistemologies of mastery and the ecological imaginary, in Rawes, P. (ed.) *Relational Architectural Ecologies: Architecture, Nature and Subjectivity*. London: Routledge.

Corby, T. (2016) *Blood and Bones: Metastasising Culture*. Available at: www.bloodandbones.org (Accessed: 1 July 2022).

Denes, A. (2008) 'Book of Dust (1972–88)', in Ottmann, K. (ed.), *The Human Argument: The Writings of Agnes Denes*, Putnam, Connecticut: Spring Publications.

Gilroy, P. (1993) The Black Atlantic: Modernity and Double Consciousness. London and New York: Verso.

Grenfell Tower Inquiry (2022). Available at: https://www.grenfelltowerinquiry.org.uk/news/closing-remarks-chair (Accessed: 23 July 2022).

Haraway, D. (2016) *Staying with the Trouble: Making Kin in the Chthulucene.* Durham CA.: Duke University Press.

Haraway, D. (2008) *When Species Meet.* Minneapolis: University of Minnesota Press.

Haraway, D. (1997) *Modest_Witness @Second_Millennium. FemaleMan_Meets_OncoMouse: Feminism and Technoscience*. New York: Routledge.

Horton, H. (2022) 'Small minority of UK parliamentarians attend emergency climate briefing', The Guardian, 11 July. Available at: https://www.theguardian.com/environment/2022/jul/11/fewer-than-10-percent-of-uk-mps-sign-up-for-emergency-climate-briefing (Accessed: 23 July 2022).

Howie, G., 2016. How to Think about Death: Living with Dying. In V. Browne & D. Whistler (Authors), *On the Feminist Philosophy of Gillian Howie: Materialism and Mortality* (pp. 131–144). London: Bloomsbury Academic.

Kolbert, E. (2014) *The Sixth Extinction: An Unnatural History*. New York: Picador.

Lewis, S. and Maslin, M. (2018). *The Human Planet: How We Created the Anthropocene*. London: Pelican Books.

Matthews, D., Matthews, D., Randers, J., Behrens, W. (1972) *The Limits to Growth: A Report for the Club of Rome's Project on the Predicament of Mankind.* London and Sydney: Pan Books.

Meadows, D.H., Meadows, D.L., Randers, J. and Behrens III, W.W., 1972. *The Limits to Growth: A Report for the Club of Rome's Project on the Predicament of Mankind*, New York: Universe Books.

McKittrick, K. (2021) *Dear Science and Other Stories*. Durham CA: Duke University Press.

Milman, O. (2022) 'US supreme court rules against EPA and hobbles government power to limit harmful emissions', *The Guardian*, 30 June. Available at: https://www.theguardian.com/us-news/2022/jun/30/us-supreme-court-ruling-restricts-federal-power-greenhouse-gas-emissions (23 July 2022).

Rawes, P. (2021) 'Visualising uncertainty and vulnerability', *Materia Arquitectura*, 20. Santiago, Chile: Université San Sebastiano.

Rawes, P. (2008) *Space, Geometry and Aesthetics: Through Kant and Towards Deleuze*. Basingstoke: Palgrave Macmillan.

Smith, S. (2022) 'Government's net zero climate strategy "unlawful" as heatwave temperatures soar', *The Independent*, 18 July. Available at: https://www.independent.co.uk/climate-change/news/net-zero-unlawful-climate-crisis-b2125648.html (Accessed: 23 July 2022).

UK Government (2022) 'Ethnicity and COVID-19'. Available at: https://www.ethnicity-facts-figures.service.gov.uk/covid-19 (Accessed: 23 July 2022).

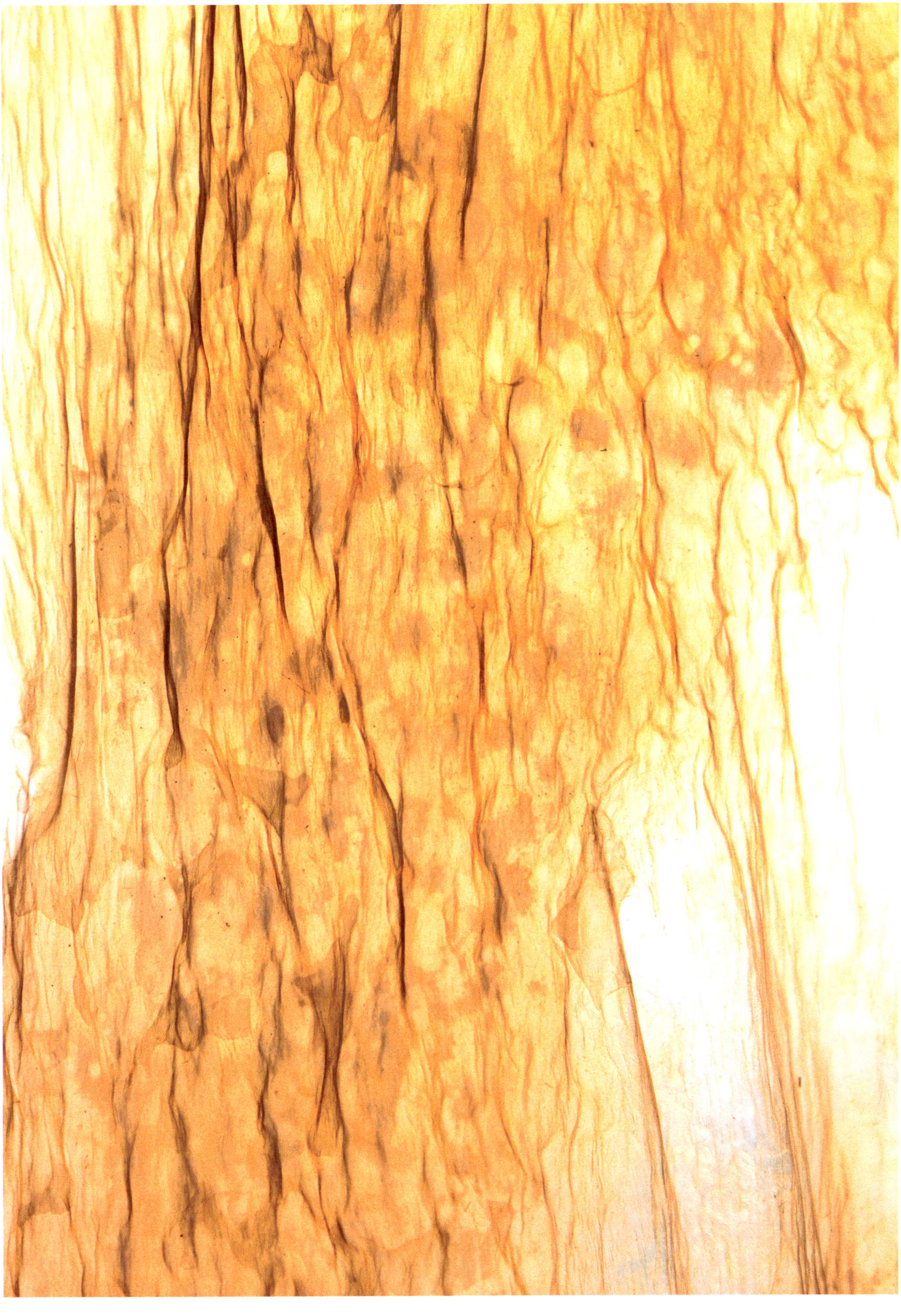

Ruth Morrow
Ben Bridgens
Louise Mackenzie

# 2

# MacroArchitectures
## Biotechnological Prototypes at the Building Scale

↗ **Glossary**
Prototypes
SCOBY

**Chapter 1.1**
↗ Page 18

**Chapter 2.1**
↗ Page 60

The concept of a MacroArchitecture has a specific relevance when working with biological materials in the built environment. Many of the materials and processes described in this book begin at a magnitude so small that they can only be described as micro: studied and acted on through the tools and processes of microbiology. The six chapters in this section each set out a transition from micro to macro – the movement from lab to architectural-scale application. Large-scale prototyping is carried out in and around the OME, | Figure 1↗ | HBBE's experimental building described in **Chapter 1.1.** Not only do the **prototypes**↗ described here develop biological materials and processes at scale, they also place biological materials and systems in context and observe and reflect upon the situated nature of how these materials and systems are in the environment. Each MacroArchitecture was born from the shared ideas of practitioners who operate across diverse and often interdisciplinary practices, from architecture and design to electrochemistry and molecular biology. Thus they approach architecture through processes of collaboration and negotiation, both with team members and with living materials and systems.

**Chapter 2.1** approaches the concept of biological architecture from the perspective of form. Beginning with only 'soft' materials, BioKnit asks how we might imagine a future in which buildings are grown. Drawing inspiration from the natural forms in the work of Gaudi, this MacroArchitecture challenges approaches to both design and assembly and begins to contend with the complexities of developing biological structures on a large scale, outside the context of the laboratory.

Chapter 2.2
↗Page 76

These challenges are further articulated in **Chapter 2.2**, which rethinks the production of bacterial cellulose for external applications in the built environment. Exploring both traditional and non-traditional methods of bacterial cellulose production that use symbiotic cultures of bacteria and yeast (or **SCOBY**↗), the researchers produce bacterial cellulose in various forms as an exterior cladding material. The resultant façade is observed in context across winter, spring and summer, by a wide range of audiences. From this a picture emerges of how we might produce, apply, maintain and live with biological matters.

Chapter 2.3
↗Page 90

The **third chapter** in this section describes concrete façade panels that repurpose existing materials in order to support life, at the macro scale for plants and insects and at the micro scale creating habitats for microbes. Biocellular concrete uses the by-products of industrial processes to create a porous material that is 90 per cent waste. The material is then formed in a manner that can sustain local plants and microorganisms such that both the material itself and the species within it are capable of absorbing and capturing atmospheric $CO_2$.

Chapter 2.4
↗Page 102

**Chapter 2.4** works with biological processes to consider how we might rethink building conservation. The Healing Masonry project reveals the potential of introducing bacteria to enable traditional construction materials to heal themselves. The project considers potential applications of bacteria that can calcify, from those found naturally in the environment to engineered strains that might improve building strength, sequester carbon, or use pigment to signal environmental pollutants. By creating a bacterial healing process which leaves a visible scar on the building façade, this biotechnology reveals and records the passing of time and the history of the material.

Chapter 2.5
↗Page 116

The **penultimate chapter** in this section takes a design approach to biological architecture by developing a materials library. BioMateriOME is a physical and experiential resource that aims to present a growing range of biological materials; for architects, designers and wider audiences. The library serves two purposes: firstly, the sharing knowledge of the origins and intended purpose of biological materials and sec-

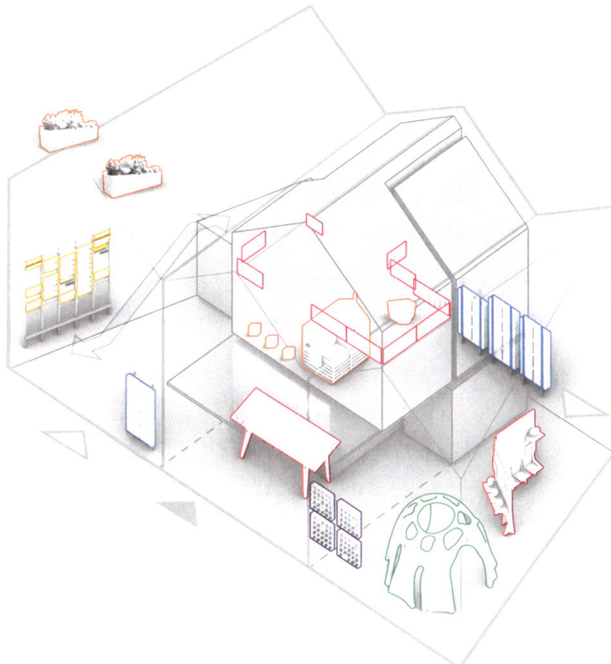

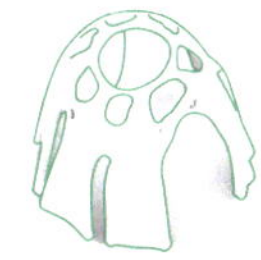

**Chapter 2.1**
Bioknit

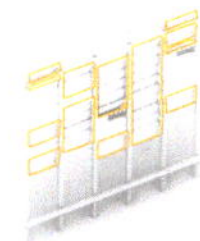

**Chapter 2.2**
Bacterial Cellulose Façade

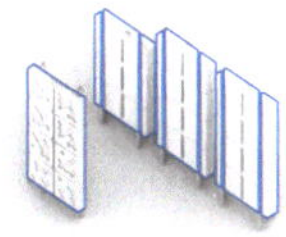

**Chapter 2.3**
Biocellular Concrete

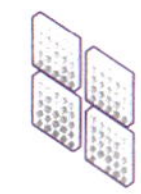

**Chapter 2.4**
Healing Masonry

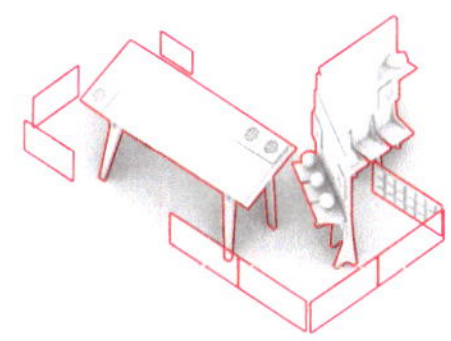

**Chapter 2.5**
Biomateriome

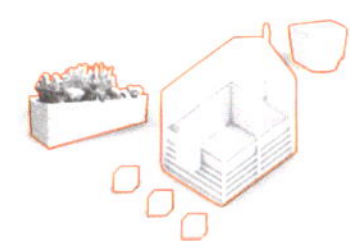

**Chapter 2.6**
Energising Waste

**Figure 1**
Locations of the Architectural Prototypes in the OME

ondly, by presenting the library as a physical materials bank in a public space, data on the microbial inhabitants of the materials can be gathered and analysed over time, shaping our understanding of the microbial ecology of materials.

**Chapter 2.6**
↗Page 130

Finally, **Chapter 2.6** imagines the self-sufficient home of the future, developing increasingly complex cycles of energy, water and nutrients in a domestic context. The starting point is three key enabling technologies: a separating toilet, microbial fuel cells and anaerobic digestion, which enable human waste to be converted into energy and nutrient-rich fertiliser. The system is developed into include food production and low-energy, traditional food preparation techniques such as fermentation, to make the best use of 'tiny energy' and convert waste streams to circular flows. The considerable cultural shift required to establish such a system in societies that rely on global food infrastructures and abundant fossil fuels is acknowledged.

Each MacroArchitecture is the result of intense collaborative and interdisciplinary teamwork. All team members contributed their expertise in concept, design, materials development and biofabrication strategy. The details indicated in the Team Lists for each chapter are an indication of specialist expertise rather than specific contributions.

# 2.1

# 2.1

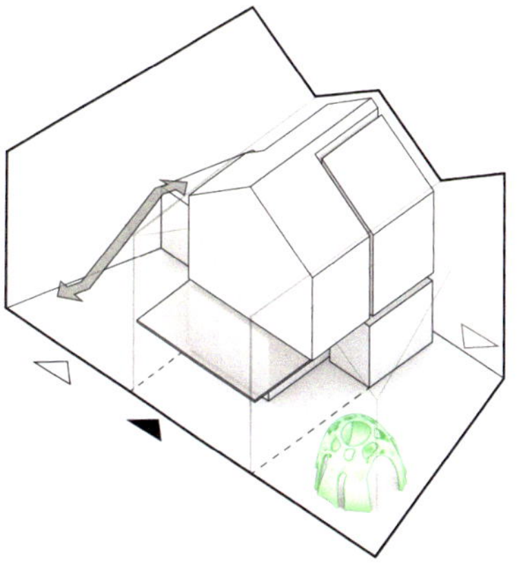

**Figure 2**
Location of BioKnit prototype in the OME

# Hi-tech / Low-tech / Bio-tech

## Crafting the BioKnit Prototype

**Data**
Application: Freestanding structure
Materials: Wool, mycelium, bacterial cellulose, sawdust, water
Dimensions of prototype: Height 1.8 m, base diameter 2 m
Weight: Wet 200 kg, dry 50 kg (approx.)

**Team**
Jane Scott: Responsive and Biohybrid Textiles, Knitting • Elise Elsacker: Mycelium Composites, Living Mycelium Materials • Romy Kaiser: Textiles, Biofabrication and Mycelium Composites • Armand Agraviador: Environmental Architect, Computational Design • Aileen Hoenerloh: Bacterial Cellulose • Ben Bridgens: Structural and Architectural Design • Ahmet Topcu: Functionally Graded Mycelium • Dilan Ozkan: Mycelium Growth and Digital Fabrication • Emily Birch: Biological Hygromorphs • Oliver Perry: Technical Officer

↗ **Glossary**
biohybrid

What could a biological architecture look like? What would it feel like to inhabit a building that has been grown rather than constructed? BioKnit explores how knitted fabric can be used in a **biohybrid** ↗ system with fungi and bacteria to produce a large-scale, freestanding structure. In contrast to the smooth, hard, rectilinear building interiors which have become ubiquitous over the last hundred years, BioKnit examines how novel construction materials enable us to rethink the spaces and surfaces we inhabit, in this case incorporating organic forms and tactile surfaces. The prototype is composed of mycelium (the root network of fungi), cellulose (produced by bacteria), and knitting (produced using 3D knit technologies). These materials have a considerably lower environmental impact compared to conventional construction materials and provide the opportunity to radically rethink how and what we build.

**Figure 1**
Detail of BioKnit illustrating mycelium (white) growing through knitted fabric (navy)

## Hi-tech / Low-tech / Bio-tech

↗ **Glossary**
mycelium
substrate
bacterial cellulose
biomaterials
biofabrication

BioKnit employs advanced knitting technologies to fabricate a shaped tubular textile that acts as a scaffold and scaling agent to guide the growth of **mycelium ↗**. Together, the knit, mycelium and bulk **substrate ↗** materials form a bespoke composite. This structure provides the framework to host two-metre-long **bacterial cellulose ↗** panels pre-grown to custom shapes providing a tactile, translucent skin that contrasts with the knitted panels. The making process uses hand assembly techniques throughout, relying on the skill and dexterity of the team to carefully bring together each element of the biohybrid system. Whilst hi-tech computation and digital fabrication tools allow for the concept of a bio-tech architecture to be realised, BioKnit underlines the importance of a tacit understanding of the materiality of each component. It brings together the hi-tech, the low-tech and the bio-tech, applying a deep knowledge of the craft of working with living materials and knitted fabrics to successfully use growth to shape future architectural construction.

BioKnit emerged from questions about how the properties of knit could be harnessed to create flexible, adaptable, low-environmental impact, tactile buildings. We were interested in exploring how the relationship between fibres, yarns and fabrics, fundamental to knitted fabrics, could be engineered to enhance the growth of **biomaterials ↗**, and how these techniques could be used to begin to think through not only how bio-tech could replace the materials with which we build, but fundamentally change how and what we 'build'.

A textile acting as a scaffold for mycelium and bacterial cellulose offers not only a technical solution for **biofabrication ↗** at a large scale, but also offers an alternative vision for the built environment. This biohybrid textile architecture has the potential to replace hard, rectilinear buildings with soft curvaceous structures and interiors, and indeed could blur the distinction between structure and interior surface. In using biological growth for construction, knit could in some places be subsumed by the living organism, losing stitch definition and changing the properties to form a rigid composite. In other places, it could retain the fabric's qualities of stretch and drape. The aim from the start was to integrate multifunctionality – diverse properties and characteristics – into a single material; qualities made possible by the combination of industrial knitting technologies, digital programming and biotechnology.

## Knitted Architecture

↗ **Glossary**
polymer

Knitting and architecture might at first seem to be an unlikely pairing: knitting is soft, flexible and extensible, in opposition to the rigid durability of architectural construction. But there is a growing interest in exploring space from the perspective of a material that stretches, expands and readily deforms to enclose objects and space. Researchers including Jenny Sabin, Mette Ramsgaard Thomsen and Sean Ahlquist have pioneered techniques to produce large-scale knitted architectures using the extensibility and tensile strength of knitted fabrics to explore space in new engaging and interactive forms. *Lumen* (Sabin Studio, 2017) demonstrates the scale achievable using knitting technologies for the MOMA PS1 series in New York. At an architectural scale, knit is also used as permanent formwork for installations such as *KnitCandela* (Popescu, 2018). Sean Ahlquist's Social *Sensory Architectures* utilise knitted fabric's ability to stretch and contract to produce touch-responsive surfaces as inclusive architectural environments (Ahlquist, 2016). These are tactile architectures that invite inhabitants to touch and interact with the structures.

The ability to produce architectural-scale knitted structures has been facilitated by advances in computation and digital fabrication, providing the tools to model the fabric's behaviour and resultant geometry prior to knitting. The latest generation of industrial knitting technologies can produce highly complex fabric structures and forms, thanks to advanced programming and sophisticated machine controls. | **Figure 3 ↗** | Programmable Knitting, developed in previous work by Jane Scott, uses industrial knitting technologies to engineer environmentally responsive shape-change behaviour into knitted fabrics (Scott, 2018). In this work the configuration of knitted stitches controls the types of shape-change produced, exploiting the potential to specify individual stitches and use different yarns on a stitch-by-stitch basis throughout a fabric.

**Figure 3**
An industrial knitting machine showing the needle beds and yarn feeders. This technology has the flexibility to integrate multiple yarns and stitch types into shaped and 3D fabrics

Alongside conventional knitted fabric production, composite textiles offer a way to bring together multiple, often opposing, properties to produce technical fabrics suitable for a variety of industrial applications. Textiles are strong and lightweight and so they are often incorporated within composites to reduce the weight of a material whilst achieving high levels of strength. Textile composites are integrated into cars and aeroplanes, as well as the fabric of buildings, because of their high strength-to-weight ratio. A textile composite usually consists of woven fibres (such as glass, carbon or aramid) encased within a **polymer ↗** resin (such as epoxy or polyester). These materials offer exceptional mechanical properties, but with negative environmental impacts including high embodied energy and limited end-of-life options. The opportunity to use biological materials in composites could offer a sustainable alternative.

## Biohybrid Construction

**↗ Glossary**
biocompatibility
fermentation

Biohybrid construction systems offer the potential to dramatically reduce the impact of the materials and processes used to construct our built environment. Working with natural fibres and microorganisms, including bacteria and fungal mycelium, could provide a more sustainable building sector by moving away from carbon-intensive materials such as steel and cement. Biomaterials also have the potential to be recycled at end of life (composted and/or shredded and used as a substrate for growing new materials), and have benefits for occupant comfort as they absorb moisture in the air and regulate internal humidity.

Mycelium, the root network of fungi, has enormous potential in architecture as a binder to create bulk composite materials from a wide range of bio-based substrates including woodchip, waste coffee grounds and agricultural waste. The resulting materials have good compressive strength so can be used as lightweight bricks, or as panels that provide a range of functions including thermal and acoustic insulation (Elsacker, 2021). In a natural environment, mycelial networks are expansive, branching out through soils to decompose organic and even inorganic materials (Stamets, 2005). To produce a mycelium composite, the conditions required by the organism are replicated within a controlled environment. Moisture, nutrients, temperature, and light levels must be carefully managed to achieve optimum growth. As mycelium consumes nutrients, it binds to the substrate materials and, when growth is stopped through heat treatment or dehydration, the end result is a coherent composite.

Mycelium has good **biocompatibility ↗** with textiles. The natural plant or animal fibres used to create a knitted fabric are composed of materials that mycelium consumes in a natural environment. Therefore depending on the constituent fibres, fabrics can provide both a scaffold to contain and direct growth, and a substrate for mycelium to consume. Tubular knitted fabrics can be filled with a mix of mycelium and substrate materials. This enables the knitting to shape the composite, as well as introducing tensile strength to improve the performance of the resulting biohybrid composite. | **Figure 4 ↗** |

In addition to mycelium, bacterial cellulose was applied to the surface of the knitted structure as a skin. Bacterial cellulose is a flexible sheet material grown through a process of **fermentation ↗**, with completely different properties to mycelium | **Figure 5 ↗** |. Bacterial cellulose can be grown either as a pure culture or through a Symbiotic Culture Of Bacteria and Yeast (known as a SCOBY). When dried, this biomaterial transforms into a leather-like translucent skin. It has textile qualities that have predominantly been exploited in clothing and product design. For BioKnit, the team wanted to explore how mycelium and bacterial cellulose could be applied to the same knit biohybrid, and how working with different microorganisms would change the materiality of the structure.

**Crafting BioKnit with High-tech and Bio-tech Processes**

**↗ Glossary**
morphologies
parametric
catenary
biocomposite

## 1 Design Concept Development

The original design concept for the BioKnit prototype was developed based on several aims:

- Demonstrate how a biohybrid textile could produce a large-scale, self-supporting structure; large enough for a person to sit inside, and incorporating some kind of skin to provide enclosure.
- Minimise material use by understanding the mechanical and aesthetic properties of each material.
- Consider how the materials function as composites to produce an efficient structure.
- Use textile formwork to hold the substrate securely and uniformly.
- Simplify the production process at all stages to reduce fabrication time. | **Figure 6 ↗** |

Inspiration for the form began with consideration of how nature builds with fibrous structures. The team looked to trees, climbing plants, fungi, nests and cocoons to create a directory of bioinspired **morphologies ↗**. Preliminary form studies were generated with computational design software using a randomised tubular geometry | **Figures 6, 7 ↗** |. **Parametric ↗** inputs allowed variation in the concentration and diameter of the tubes and this led to a variety of initial design ideas.

The use of computational design tools were particularly valuable for exploring material properties both during and after growth. The design process needed to consider the materials in different states to account for the requirements of biofabrication. Materials were both wet and dry, and soft and stiff, at different stages of the process. The initial concern was that the knitted structure would stretch to a high degree when it was filled with a heavy, wet mycelium substrate mix. Both the scale of the structure and the lack of rigidity of the knit and mycelium during the growing period were key concerns. Computational modelling enabled the team to predict how the structure would behave using physics simulation, illustrating the impact of gravity on textile tubes containing different substrate weights.

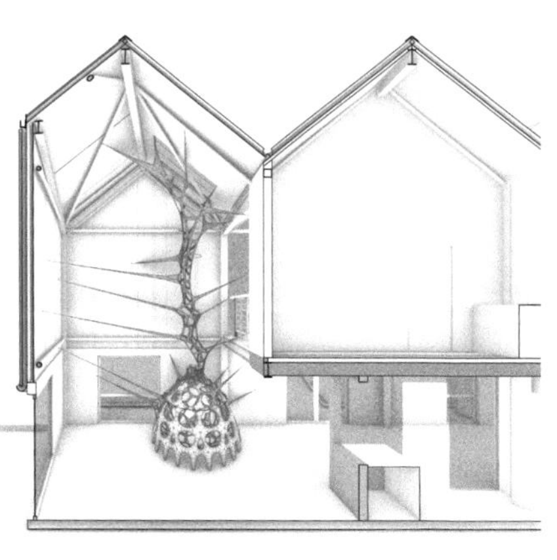

**Figure 6**
3D conceptualisation of BioKnit in the OME

Figure 4

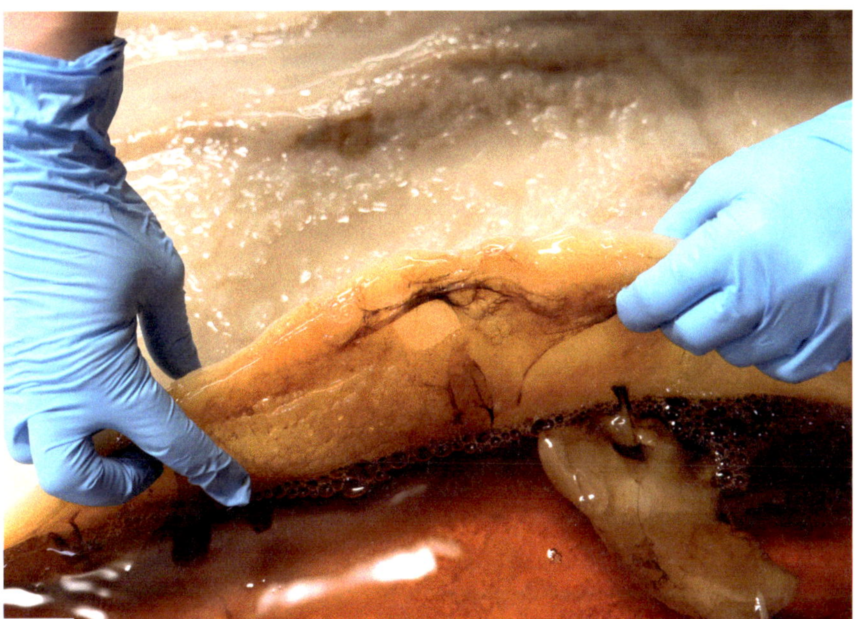

Figure 5

**Figure 4**
Mycelium/knit composite illustrating the mycelium (white) growing through the substrate (wood fibres) and tubular knitted fabric (black)

**Figure 5**
Detail illustrating bacterial cellulose growing

**Figure 7**
Bioinspired sketch catalogue

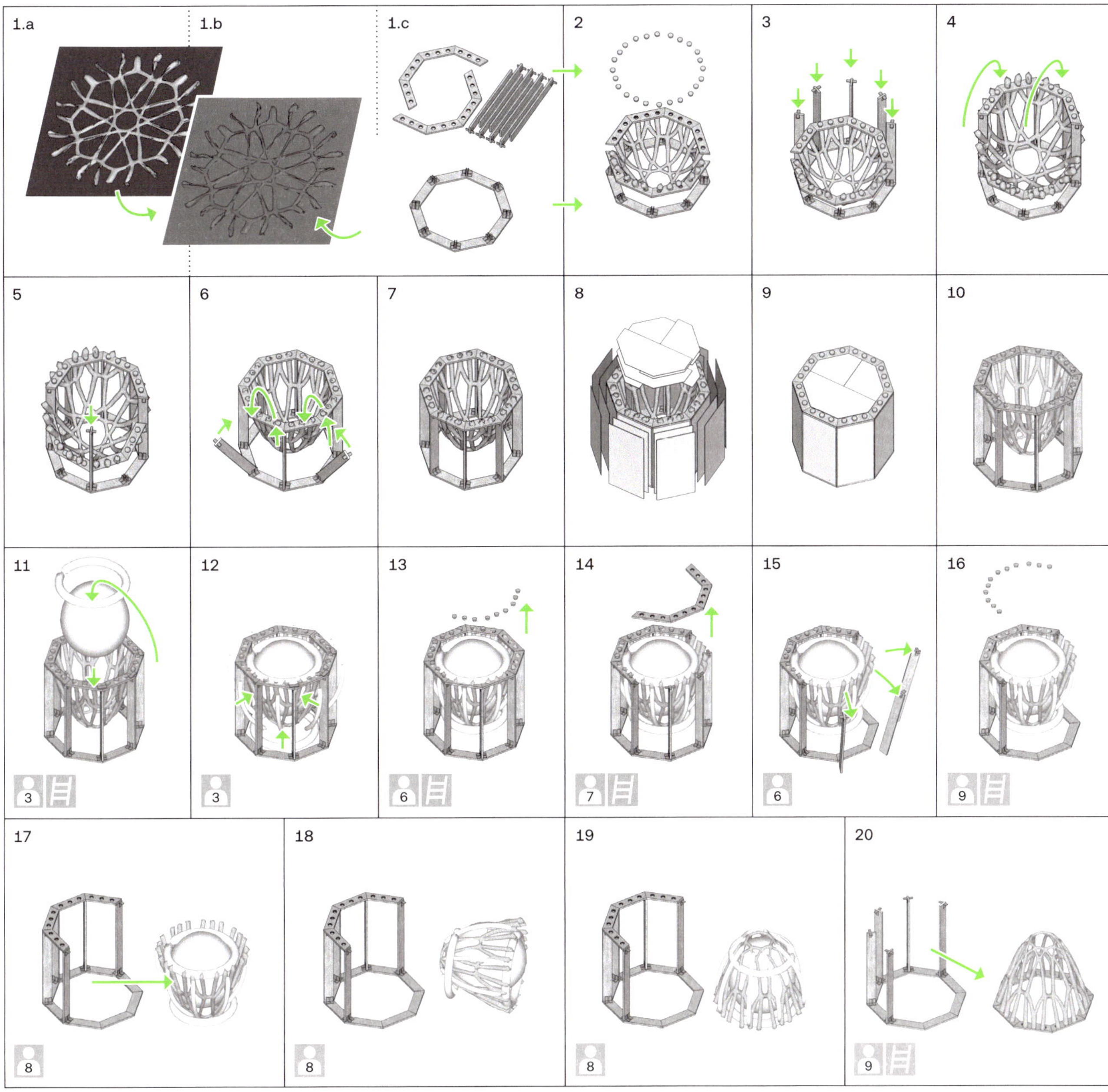


**Figure 8**
Storyboard illustrating assembly of the growth chamber

**Suspension (1 – 9)**

1.a Fill knit over liner
1.b Slide base under knit
1.c Move remaining timbers to OME
2 Position upper ring halves at workable height (use stools) Pull or tube ends through corresponding hole and secure
3 Insert and secure 5 columns
4 Raise first half of upper ring onto columns and secure (second half may need to be raised slightly onto higher temporary supports depending on cross-tube length)
5 Tilt remaining half-ring to insert and secure 1 column (inserting all remaining columns not possible until half-ring is clear and above)
6 Raise upper ring second half onto columns. Insert final 2 columns and secure all with screws
7 Raise second ring
8 Triple layer foil (staple)
9 Sealed Chamber, remember sensor inside

**Inversion (10 – 20)**

10 Matured Basket
11 Pack basket with large balloon and inflatable tube as ring
12 Use remaining inflatable tubes to pad basket exterior and secure gently with straps
13 Remove staples around 9/21 corks or cut around fabric. Team to support 9 legs.
14 Remove 3 sides of upper ring. May need saw for biscuit joints.
15 Remove 3 columns
16 Detach remaining legs from corks. Full team to support all legs.
17 Basket moved out of frame
18 Basket rotated, careful to avoid sag or breakage at 90 degrees.
19 180 degrees to dome
20 Inflatable packing/padding removed and frame disassembled. Top ring can be used to brace bottom of dome if required.

A major step in the design development resulted from combining two pragmatic solutions. Firstly, the structure was designed from a series of structurally identical, shaped modules that could be knitted individually and then assembled to form the complete form, prior to filling and inoculation with mycelium. The second solution was a decision to use the soft-to-hard phase change, that is, the point at which the wet living structure dries out, to enable a structurally optimal form to be achieved.

The idea of using the transition from flexible to stiff via growth enabled the design team to optimise the structure of the biohybrid using **catenary ↗** logic. Inverted catenary curves have been utilised in arches, vaults and domes throughout architectural history due to their unique ability to carry their own weight, and any uniformly distributed loading, in pure compression. The concept for BioKnit was to suspend the prototype in tension in a soft state, and through the process of growth, form a catenary dome that could be inverted once the composite had hardened. This process takes advantage of the soft material phase necessary for the growth of a biohybrid textile structure, exploiting the drape and tensile strength of the knitted fabric during hanging, resulting in a self-organising, form-making technique for **biocomposite ↗** construction.

During production, models designed using 3D modelling software generated the specifications for the geometry of the repeated knitted module, as well as estimations of the volume of mycelium substrate required to fill the stretched knitted tubes. Each stage of fabrication was also planned visually, troubleshooting some of the most complicated processes that were developed for this unique material system through detailed storyboards and scale models | **Figure 8 ↗** |.

## 2 Mycelium + Knit

Early material experimentation focused on the biocompatibility of each microorganism with knitted fabric. Initial mycelium + knitting experiments tested the growth of different mycelium species within a tube of knitted fabric, exploring the most successful yarns and fabric structures to promote mycelium growth, and what mix of substrates were appropriate when working with flexible knitted scaffolds | **Figure 9 ↗** |. Three different mycelium species were used for initial tests: *Pleurotus ostreatus*, *Trametes versicolor* and *Ganoderma lucidum*. These species were conserved on a grain mixture, called spawn, at 4 °C and combined with a mix of nutrients and bulk materials prior to packing into the knitted tubes. This is known as the mycelium substrate mix. The mycelium knit composites were grown in darkness at a constant temperature and humidity for two weeks. Whilst all species grew successfully, *Ganoderma lucidum* was selected based on the compressive strength of this mycelium as a composite. Results of early experimentation identified that all natural fibres had excellent biocompatibility; wool was selected for use in tubular structures and linen was selected for contrast within fabric panels. The fabric structure did impact growth: looser, more open structures benefitted from quicker growth; however, the use of these open structures led to uneven distribution of the mycelium substrate.

As the research progressed, the fabric density became an important factor; a tight fabric composed of multiple yarns (thicker than standard for the needle size in the knitting machine) was selected to control the large volume of wet mycelium substrate. The recipe for the mycelium substrate mix was developed to provide a more viscous substance that could be packed uniformly into long, narrow knitted tubes.

## 3 Bacterial Cellulose + Knit

Bacterial cellulose is produced in a liquid media at the air-liquid interface where the bacteria have access to nutrients and oxygen simultaneously. The material forms a sheet that fills the surface of the container it is grown in. Early experiments tested how to produce a composite of bacterial cellulose and textiles by placing

Day 6 Day 12 Day 20

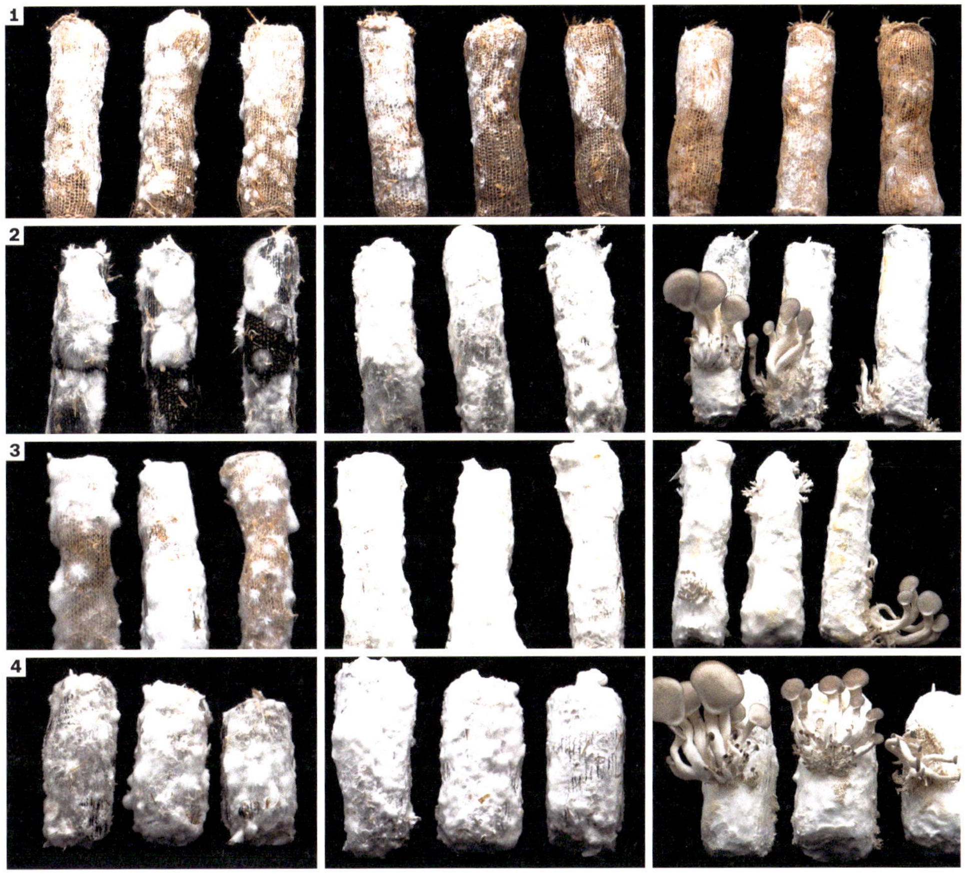


Day 6 Day 12 Day 20

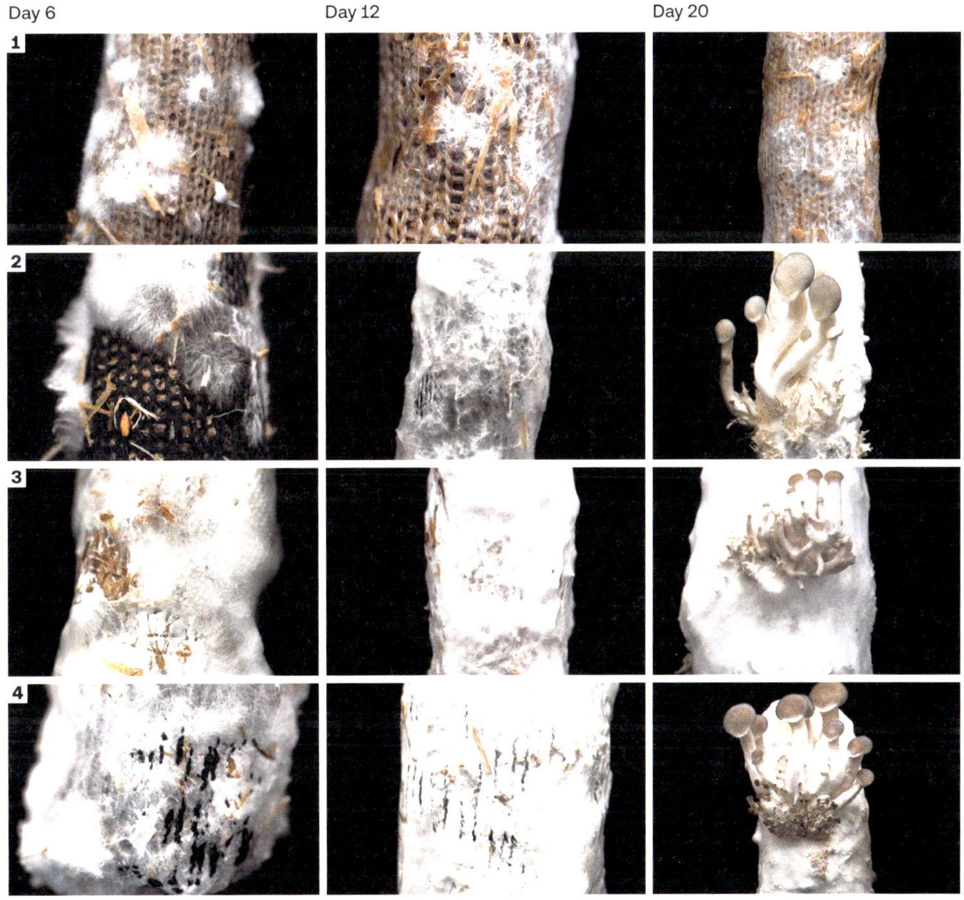


**Figure 9**
Mycelium + Knit fibre and yarn tests

the textile at the air-liquid interface so the bacterial cellulose could grow through it. These experiments were carried out with a variety of knitted fabrics to understand the attachment strength achieved between the two materials in a wet and dry state. The attachment was the strongest on fabrics with larger yarns and a more open knit structure. Here, the bacterial cellulose grew partly through the openings and formed strong connections between the two layers underneath and on top of the fabric. This early experimentation helped to inform the strategy to integrate bacterial cellulose and mycelium using a textile interface.

## 4 Mycelium + Bacterial Cellulose + Knit

For the final prototype the ambition was to integrate both biomaterials within the same biohybrid, but as we have seen mycelium and bacterial cellulose require very different environmental conditions to grow, and so combining them presented a challenge. Tests were undertaken to assess the most suitable way to bring the different materials together (Hoenerloh et al., 2022). Ultimately, a decision was taken to grow the two organisms separately and combine them during the final drying stage | Figure 10 ↗ |. The bacterial cellulose dried onto the mycelium knit composite with good attachment, especially in areas where the mycelium had created a white film on the fabric. This provided the most efficient method of scaling up with the lowest risk of contamination.

### Growing a Lightweight Structure

The BioKnit prototype stands at 1.8 m high with a diameter at the base of 2 m. It was grown as a single piece, in situ, in the OME. During growth the knitted form itself weighed around 15 kg and contained approximately 185 kg of mycelium substrate. The whole prototype was saturated with water to enable the mycelium to grow. In fact, water accounted for about 75 per cent of the weight during the growing stage. During the growth period, mycelium hyphae extended outwards throughout the substrate mix and the knitted fabric to consume available nutrients and bind the materials together. As the structure air-dried and the water evaporated, the weight decreased significantly. The final structure weighs approximately 50 kg.

During growth, BioKnit was suspended within a bespoke growth chamber to achieve catenary arches and provide the environmental control necessary for mycelium growth | Figure 11, 12, 13 ↗ |. Mycelium composites are conventionally grown in lab conditions where temperature, light levels and humidity can be regulated. This was not feasible for BioKnit due to the scale of the project growing as a single piece. The growth chamber therefore replicated the lab conditions as closely as possible. Temperature was maintained between 21 and 28 °C, and relative humidity maintained at 50 per cent. Thermal reflective foil covered the growth chamber to provide darkness. The mycelium grew inside the growth chamber for 13 days, after which the foil layers were removed and the structure was left to air dry for another 22 days.

The dried structure was then rotated 180 °C to form the catenary vault. After being inverted, the structure became freestanding. By combining the compressive strength of the mycelium composite with the tensile strength provided by the knitted fabric, the prototype was structurally very strong and able to withstand an additional 35 kg weight of bacterial cellulose panels (saturated weight) applied to the surface of the structure. On drying, the bacterial cellulose provided a translucent skin.

### Knitting as a Scaffold for Biofabrication

The BioKnit prototype exploits both the formability and high strength-to-weight ratio of a knitted textile. The fabric is transformed into a self-supporting structure by the rigidity of mycelium composite and the structural efficiency of catenary geometry | Figure 14 ↗ |. A key success of the making process is the creation of a large-scale biohybrid structure in a single piece. The slender tubes that support the structure form a continuous network that stretches throughout the prototype, providing the

Figure 10

Figure 11

**Figure 10**
Bacterial Cellulose Coating
Mycelium Knit Composite

**Figure 11**
Hand filling the Knitted preform
with mycelium paste. Care is
required to keep all surfaces clean

Figure 12

Figure 14

Figure 13

**Figure 12**
Manipulating the filled knit preform so that the prototype legs are suspended in the roof of what will become the growth chamber

**Figure 13**
BioKnit Suspended in Growth Chamber

**Figure 14**
BioKnit on Exhibition in the OME

**Figure 15**
Detail showing knitted fabric with bacterial cellulose skin filtering light

necessary support without additional connections or support systems. In fact, growth becomes a way to bring together a series of complementary materials – wool, linen, mycelium and cellulose – in a continuous intertwined structure. BioKnit demonstrates the resilience of growth as a composite forming method: contamination did not impact the structural performance of the prototype. Minor contamination that did occur – as conditions were not sterile – was removed manually to prevent potential problems.

## Making and Growing as an Interdisciplinary Design Process

The design, fabrication and growth of BioKnit presents a model for interdisciplinary practice that combines a range of knowledge and skills to start to contemplate what a biological architecture could look like, how it can be designed and fabricated, and how it might behave. From a design perspective, textile experts working alongside architects were able to address materiality as well as formal and technical aspects of the design. This resulted in the shape of BioKnit being generated, at the knitting stage, as a series of seven shaped modules assembled together to make one cohesive knitted form that included tubular sections, fabric meshes and integrated tubular branches to join the modules together. Experts in mycelium and bacterial cellulose challenged their own assumptions about the materials to explore the potential for growing these materials together and outside the laboratory | **Figure 15↗** |. The team considered how, given the right growing conditions, competition between organisms could be used to change, improve and challenge how a biomaterial grows. The ambition to grow a structure large enough to occupy motivated the team to challenge the available technologies. For example, during knitting the full machine width was used alongside extensive shaping to create unique shaped pieces necessary to create the overall form. In addition, by working between textiles and biofabrication the mycelium substrate mix was optimised for use with textiles to enhance the consistency and workability in soft moulds – soft enough to enable the fabric to form during hanging.

Whilst BioKnit integrates high-tech industrial knit technology with bio-tech materials research, there is a vital element of craft in how these materials have been cultivated to form a cohesive biohybrid structure. In fact, the tacit knowledge gained through the extensive experience of growing and making across a range of materials and processes was fundamental to achieving a successful outcome, and highlights the importance of the making process in delivering innovation in bio-tech. Textile thinking in particular offers a materials and process-led methodology that benefits the interdisciplinary way of working central to success within this field. This fabrication system finds its way back to principles of making and demonstrates why textiles, a discipline that sits between craft and technology, is such a valuable approach with the potential to transform biofabrication for the built environment.

## References


Ahlquist, S. (2016) 'Sensory material architectures: Concepts and methodologies for spatial tectonics and tactile responsivity in knitted textile hybrid structures', *International Journal of Architectural Computing* 14(1), 63–82.

Elsacker, E. (2021) 'Mycelium matters: An interdisciplinary exploration of the fabrication and properties of mycelium-based materials'. PhD thesis, VUBPRESS.

Hoenerloh, A., Ozkan, D., Scott, J. (2022) 'Multi-Organism Composites: Combined Growth Potential of Mycelium and Bacterial Cellulose', *Biomimetics* 2022, 7, 55.

Popescu, M. et al. (2018) 'Building in Concrete with an Ultra-lightweight Knitted Stay-in-place Formwork: Prototype of a Concrete Shell Bridge', *Structures* 14(3), 322–332.

Sabin, J., Pranger, D., Blinkley, C., Strobel, K., and Liu, J. (2018) 'Lumen', Proceedings of ACADIA 2018, Mexico City, 444–445.

Scott, J. (2018) *Responsive Knit: the Evolution of a Programmable Material System*, in C. Storni, K. Leahy, M. McMahon, P. Lloyd, and E. Bohemia (eds.), Proceedings of DRS2018, Vol. 4. Limerick, Ireland: Design Research Society. 1800–1811.

Stamets, P. (2005) *Mycelium Running*, Ten Speed Press, New York.

# 2.2

# 2.2

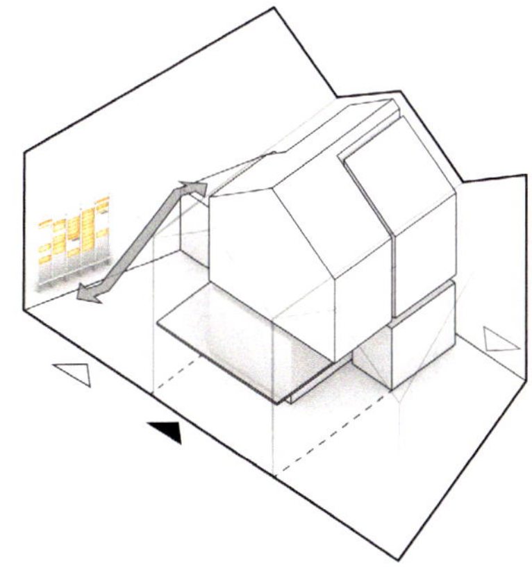

**Figure 2**
Location of Bacterial Cellulose Prototype at the OME

# Bacterial Cellulose
## Growing a Kombucha-shingled Façade

**Data**
Application: Façade prototype and other supporting experiments
Materials: Bacterial Cellulose (BC), metal and timber frame, stainless steel wire.
Dimensions of prototype: 2 × 2 m
Approx. size of BC panels: 127.5 × 355 mm

**Team**
Ruth Morrow: Architect, Material Developer • Karolina Bloch: Architect, Design Researcher • Mahab Aljannat: Molecular Microbiologist • Armand Agraviador: Environmental Architect, Computational Designer • Kaajal Modi: Design Researcher • Oliver Perry: Technical Officer

**↗ Glossary**
prototype
Bacterial Cellulose (BC)
SCOBY

This **prototype**↗ and its associated experiments test the viability, challenges and opportunities of using **Bacterial Cellulose (BC)**↗ grown from **SCOBY**↗ as a potential material for use in the built environment. This chapter explores the material properties of BC, shares insights from growing it in sheet form in the lab, and installing it as a façade material. It captures responses from researchers, students, visitors and passers-by, and presents potential directions for future research.

**Figure 1**
Closeup of Biosandstone shingles

## Critical Probes in Critical Contexts

↗ **Glossary**
Bacterial Cellulose

In this chapter we explore the process of testing **Bacterial Cellulose** (BC) as a highly novel, sustainable biomaterial within the built environment. Bacterial cellulose is a gelatinous mat formed by the secretions of cellulose-producing bacteria. Over the last few decades, it has been intensively researched for its applications for biomedical products and food; and many large corporations, including Goodyear and Sony, have filed patents involving BC for product development. It has also been explored by designers in the contexts of furniture, fashion, textiles, product, packaging and consumer electronics. However, BC's application in the built environment has been comparatively underexplored, limited only to trialling bacterial cellulose fibres in concrete as a means to reduce water absorption, improve mechanical properties, and serve as nano-bridges to prevent cracking (Guo et al., 2020).

The prototype described here represents the first attempt to use BC as an exterior cladding material. By installing the BC prototype onto the façade of the OME, we were immediately confronted by the challenges of an external context, where the risk of failure was high but equally there was a chance to uncover unanticipated potentials. The prototype thus became a critical probe in a critical context, where researchers and the public could observe its behaviour in the external environment.

By developing a large, 'building-scale' prototype and testing its limitations and potentials, we hoped to avoid investing resources and effort into resolving problems that might be significant at micro scale but become less so at the scale of the building. | **Figure 3** ↗ | It also located the work in the public sphere, where passers-by could interact with it. In its 'natural' state, bacterial cellulose has an unfamiliar almost alien texture, like an organism from the imagination of H. G. Wells. As a result, the material not only challenges existing forms and norms of manufacturing and construction, it challenges perceptions of what a city could and should look, feel and even smell like. In the following text we discuss how we implemented this approach, what we learned from it, its successes and limitations, and how we intend to take forward our learning into future research areas.

## Opportunities of Bacterial Cellulose

↗ **Glossary**
crystallinity
polymerisation
biocompatible

Bacterial cellulose is chemically identical to plant cellulose, forming a dense matrix of fibres established through heavy inter- and intra-locking chemical bonds. However, BC has higher **crystallinity** ↗, **polymerisation** ↗ degree, chemical purity, water absorption capacity and tensile strength than plant cellulose. BC is also UV-resistant, **biocompatible** ↗ and biodegradable (Gregory et al, 2021). In testing, the mechanical, thermal, and physiological properties of BC have been shown to be stronger than those of plant cellulose. | **Figure 4** ↗ |

Researchers estimate that the production of BC can achieve a comparable production efficiency with the growth of plant cellulose, and indeed, the company Nanollose, which produces BC yarn, suggests that the yield rate is significantly more per annum than plant-based cellulose, since it requires neither earth nor sun to sustain its growth, and hence can be grown in stacked containers. | **Figures 5** ↗ |

| Substance | Tensile Strength (MPa) |
|---|---|
| Soft woods | 70 to 140 MPa |
| Bamboo | 265 to 500 MPa |
| Bacterial Cellulose | 200 to 300 MPa |

Figure 4

| Substance | Growth Period | Tonnage Per Annum |
|---|---|---|
| Cotton | 6-8 months | 0.5 (raw cotton) |
| Trees | 12-18 years | 3.5 (cellulose pulp) |
| Bamboo | 3-7 years | 1.2 (cellulose pulp) |
| Microbial Cellulose | 10-15 days | 115 tons |
| Microbial Cellulose (stacked vertically) | 10-15 days | 1115 tons |

Figure 5

Figure 3

**Figure 3**
A side view of the BC Prototype in situ on the OME façade

**Figure 4**
Comparative table of tensile strength of cellulose-based materials

**Figure 5**
Table of growth periods and tonnage per annum across comparative sources of cellulose (derived from data on Nanulose website: nanollose.com)

## Symbiotic Culture of Bacteria and Yeasts (SCOBY)

In this prototype we worked chiefly with biomats of cellulosic fibres, also known as 'pellicles', which are formed at the liquid-air interface during a process derived from kombucha tea production. The pellicle formed during the kombucha process is often called a SCOBY, that is, a Symbiotic Culture of Bacteria and Yeast. The liquid and the pellicle generally include species of *Acetobacter* bacteria, as well as various *Saccharomyces* (a genus of fungi that includes many species of yeast) which are not present in BC cultures grown in the lab. We chose the Kombucha Method to grow SCOBY pellicles: firstly, as it is the simplest method for non-bioscientists to work with, being used in kitchens around the world to grow kombucha; and secondly, as team members with previous experience who had found SCOBY to be more resistant to contamination hypothesised that the diverse microorganisms in the culture rendered it more resilient.

## Shingle Construction

In order to create a façade using bacterial cellulose, we turned to an existing means of assembly: timber shingle construction. Shingles are an ancient method of cladding buildings where small elements of cleft timber are used to cover large surface areas. They can be used on roofs and vertical surfaces, and the means of assembly – where shingles are fixed through to raised battens – means that the air movement around them allows for cycles of wetting and drying without significant deterioration. Most shingled façades have a lifespan of thirty to forty years depending on exposure, though there are cited examples of shingled walls surviving from the thirteenth century (Berge, 2015). Shingle construction also enabled us to restrict production to growing smaller elements. BC can be grown in large bespoke containers, for example some designers have grown BC up to 2 m². We chose to use Euro boxes, a standardised container used in the food industry, most significantly in the mass production of Nata de coco, a traditional Philippine dessert comprised of microbial cellulose |**Figure 6↗**|. Working with such boxes resulted in manually manageable and relatively standardised BC shingles, avoiding future energy costs related to machinery for production and transport.

**Figure 6**
Bacterial Cellulose grown in stacked Euro boxes

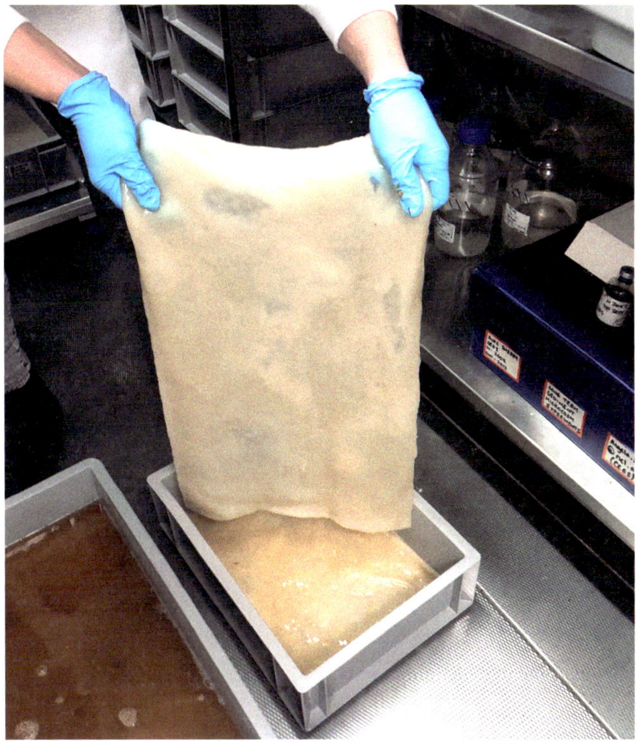

**Figure 7**
Bacterial Cellulose being harvested

## Growing BC Shingles

The standard recommended growing medium for the production of pure bacterial cellulose in laboratory conditions is Hestrin–Schramm (HS). However, cultivation of bacterial cellulose using a symbiotic culture does not require the use of HS media prepared in sterile lab conditions. Instead, BC can be grown using SCOBY pieces and a simple sugary tea medium, in the same way as kombucha tea is produced. | **Figure 6 ↗** |

For the prototype, BC production used the following ingredients.

- 1 × SCOBY pellicle (approx. 2 cm thick circle with a diameter of 15 cm, in approx. 50 ml liquid media)
- 2 litres boiled water
- 6 green tea bags (provides nutrients, minerals, caffeine for nitrogen production)
- 160 g cane sugar (is converted into cellulose by the microorganisms)
- 100 ml organic apple cider vinegar

The BC medium was prepared in batches. Green tea bags were added to boiling water and allowed to brew for a few minutes. The tea bags were removed and sugar added while the tea was still warm. The medium was allowed to cool and apple cider vinegar was added to create an acidic environment helping to reduce the risk of contamination. Thereafter the SCOBY pieces were, together with the medium, added to Euro box containers, sanitised with 70 percent ethanol.

All the boxes were closed with loose-fitting lids, stacked and allowed to ferment in a temperature-controlled room, and harvested after two to three weeks, depending on the desired thickness | **Figure 7 ↗** |. The research literature typically gives the growing temperature of lab-grown BC at 30 °C, although more recently a new bacterial strain has been developed in China that grows at 10–20 °C (Zhong, 2020). Due to space restrictions we moved pellicle production to a new location where the temperature varied, both diurnally and seasonally. We noticed no significant change in the growth of the pellicles, welcoming the reduction in energy input. Once we had harvested the first ‘crop’ of pellicles we were able to set the same media aside and use it to produce second and third crops without additional SCOBY or preparation of new media. | **Figure 9, areas E1, E2 and F ↗** |

**Figure 9**
Experimental phases of the
Bacterial Cellulose prototype

## Hanging the Shingles

We constructed a frame on the southwest façade of the OME building, using a proprietary metal framing system (Unistrut) and stainless steel wires, in order to create a 'neutral' hanging system that would neither interact with the shingles nor absorb moisture. The freshly harvested, wet BC pellicles were hung directly onto the wires, like clothes on a washing line | **Figure 8 ↗** |.

It was an intuitive and forgiving means of construction that required no fixings. At times the freshly harvested shingles were oddly shaped and inconsistent in thickness, but as they dried, they bonded to one another and shrunk in place, creating a coherent vertical surface. With each harvest and application of BC shingles, another lesson was learned, triggering the next experiment. For example, a series of BC harvests gave us an opportunity to hang shingles in alternative ways, apply different finishes and trial alternative growing conditions. Figure 9 shows the prototype ten months after the first BC shingles were hung. It is labelled to correspond to the trials and experiments set out in | **Figure 10 ↗** |.

| Area on Prototype | Experiment | Notes | Period of Growth of BC | Days on Prototype To date |
|---|---|---|---|---|
| A | Growing and Hanging Stage 1 | Portrait hanging. Grown in controlled 30 °C room | 4 weeks, 4 days | 9 months |
| B | Growing and Hanging Stage 2 | Landscape hanging. Grown in controlled 30 °C room | 2 weeks, 3 days | 7 months |
| C | Growing and Hanging Stage 3 | Mix of portrait and landscape hanging. Grown in uncontrolled temp | 3 weeks | 3 months |
| D | Growing and Hanging Stage 4 | Mix of portrait and landscape hanging. Grown in uncontrolled temp | 2 weeks, 2 days | 3 months |
| E1 \| E2 | Reuse of Growing Media Stage 1 | Reusing media from D for a 2nd growth cycle | 3 weeks | 2 months |
| F | Reuse of Growing Media Stage 2 | Reusing media from D for a 3rd growth cycle | 3 weeks | 1.5 months |
| G | Liquid SCOBY Cultivation | Cultivated with fresh liquid culture only (without SCOBY pellicle) | 2 weeks, 4 days | 1.5 months |
| H | Surface Treatment 1 | Application of Linseed Nourish Wood Oil (Brouns & Co) on BC | N/A | 4.5 months |
| I | Surface Treatment 2 | Application of Linseed Paint Exterior (colours: Barn Red, Mountain Blue, Moss, Pearl Grey from Brouns & Co) on BC and pine wood **(Figure 11)** | N/A | 4.5 months |
| J | Surface Treatment 3 | Application of Linseed Paint Exterior (colour: Moss from Brouns & Co) on BC | N/A | 1.5 months |
| K1 \| K2 | Biosandstone | Applied in a variety of ways | N/A | 0.5 months |

**Figure 10**
Table of experiments

## Biosandstone

**↗ Glossary**
biomatrix

One of our Master of Architecture students, Jay Hallsworth, took on the conceptual challenge of making BC 'more like concrete'. The student's research led to the development of hybrid bio-inorganic matrices where aggregates, such as sand or other fine waste particles, were integrated at the formative stage of the BC. By carefully managing this process the aggregate becomes embedded within the matrix of the BC, creating a more robust, though still flexible, material. The final product, named Biosandstone, is textured and comes in a variety of earthen tones depending on the nature and concentration of aggregate distributed in the **biomatrix ↗**. A range of shingles, in terms of colour and texture, were produced, offering the possibility of matching to existing building materials. | **Figure 9, Area K1 and K2 ↗** |

Figure 12

Figure 8

Figure 11

Figure 12

**Figure 8**
Application of Bacterial Cellulose shingles hung in different orientations on the OME façade

**Figure 11**
Close-up of Linseed Exterior Paint (Brouns & Co) on Bacterial Cellulose and pine wood in May 2022 (left) and September 2022 (right)

**Figure 12**
Close-up of various wet (left) and dry (right) Biosandstone samples

## Getting to Know the BC During its Growth Phase

Alongside the production of pellicles for the façade prototype, we also carried out lab-based experiments to better understand the BC during its growth phases.

During the production of the shingles we noticed that some areas of the pellicles were thinner, and that these areas seemed to be associated with bubbles forming under the pellicles. We investigated this in lab-based experiments. In | **Figure 13**↗ |, BC pellicles can be seen growing in four test tubes. Tubes 1 and 2 were filled with growing media and SCOBY (Kombucha Method) and tubes 3 and 4 contain only *Komagataeibacter xylinus* (*K. xylinus*) bacteria and HS medium (lab method). The SCOBY pellicles are clearly disrupted by bubble formation, whilst the *K. xylinus* pellicles show no bubbles, even after an extended incubation period. Hence, bubbles cannot be avoided when using the Kombucha Method. It's recorded anecdotally that DIYers working with SCOBY push the bubbles to the edges during growing. However to avoid this manual and potentially contaminating intervention, we trialled a range of sizes of containers. After several cycles of growing and harvesting, we settled on medium-sized containers (300 × 400 mm) where the bubbles were more likely to escape at the edges | **Figure 6**↗ |. This phenomenon of *bubble formation* requires further investigation, since one of the benefits of BC is that it can be grown to a size and shape that suits the intended purpose.

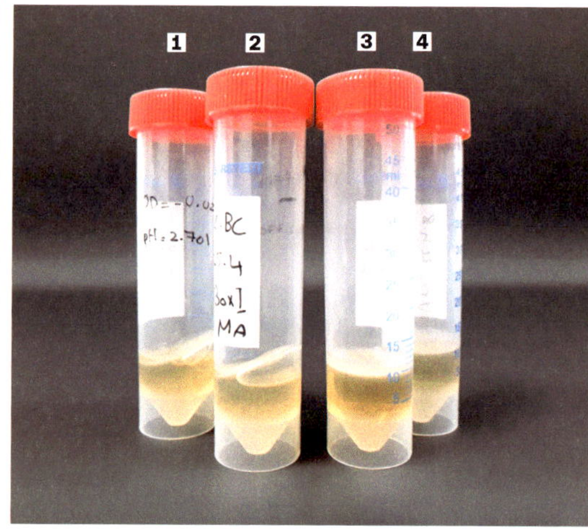


**Figure 13**
Test tubes with bacterial cellulose with SCOBY culture (tubes 1 and 2) and pure Komagataeibacter xylinus culture (tubes 3 and 4)

We also carried out simple weight tests of freshly harvested BC pellicles in comparison to pellicles dried at room temperature for three days | **Figure 14**↗ |. The average weight across three containers (Euro boxes) of freshly grown, 'fully saturated' BC pellicles was 33.53 g ± 5.11 g and the average dry weight was 2.41 g ± 0.45 g. So approximately 93 per cent of the fresh weight was liquid culture. This echoes the research literature's acknowledgement of BC's *extreme hydrophilic nature* and is clearly significant for producers when considering transport costs and carbon footprint of BC in its wet and dry states.

The Kombucha Method of growing BC for design applications typically calls for the inclusion of cider vinegar to reduce contamination from moulds and other contaminants; however, the bacteria in SCOBY also produce acetic acid. We wanted to understand this better so we monitored the pH of the media across the growth cycle and found that the pH dropped during BC formation to become increasingly more acidic, with no discernible negative impact on the growth, and no contamination | **Figure 15**↗ |.

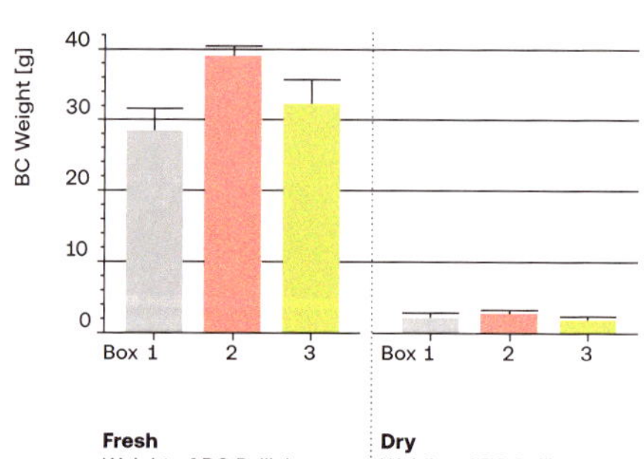


Figure 14

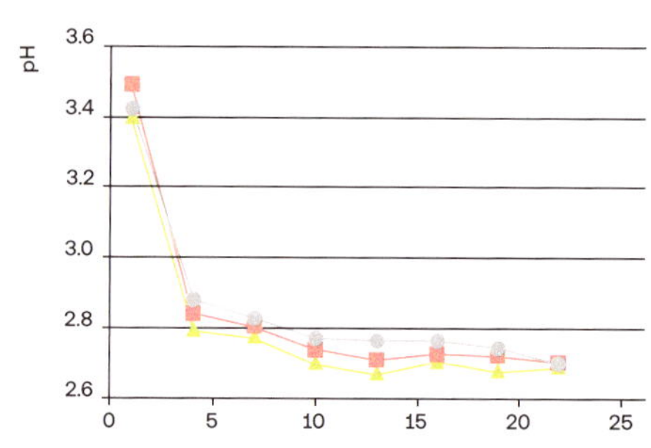


Figure 15

**Figure 14**
Comparing across three sets of samples the weight of BC pellicles before (freshly harvested; Left) and after drying (dry weight; Right) at room temperature. From 400 × 300 mm containers grown for 4 weeks

**Figure 15**
Graph showing the pH of SCOBY taking over 22 days growing time

**Figure 16**
First part of the prototype in situ on the OME façade. From the top left:
[1] Day 1
[2] Day 4
[3] Week 1
[4] Week 3
[5] Week 5
[6] Week 8

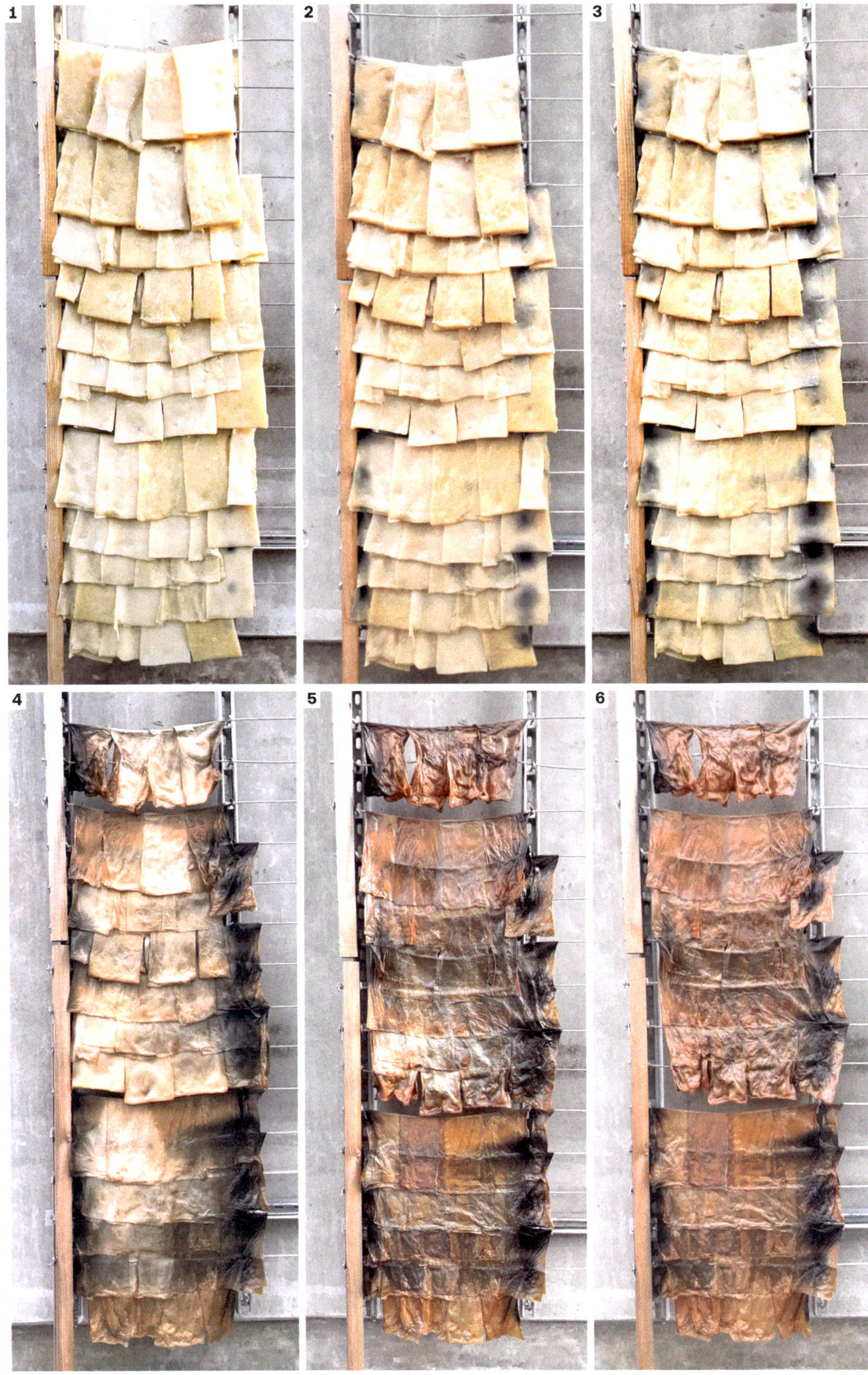

## Observations

We monitored the BC façade prototype over a ten-month period, where external temperatures varied from minus 6 to plus 37 °C, and humidity varied from 22 per cent, at its driest, to 100 percent (timeanddate.com, 1996). Once applied, in wet form, to the prototype frame, the BC shingles lose the majority of their water content within the first three weeks and thereafter dry out further but much more slowly. Ten months later they still maintain to the touch a certain moisture level (yet to be quantified through testing). The evaporation of water from the BC shingles can be noted in the colour change of the prototype, where, by week three, the creamy tones were replaced by brown tones that deepened over time. We believe this colour change is due to a process of oxidation occurring in the BC. In the early stages of the prototype, some of the fixings that held the wires in place came into contact with the BC, resulting in blackened edges. We have yet to test the degree to which this alters the BC; however, on changing to stainless steel fixings this phenomenon did not reoccur. Another notable change was the curling up and distortion of the BC at its edges. This process can also be witnessed in other materials since the edge condition is always more vulnerable to weathering.

In terms of fire resistance, BC's response to the spread of flames is currently under-researched. We carried out a very basic test, setting light to strips taken from the oldest sections of the prototype. They briefly caught alight, the moisture retained within them evaporated, and the samples charred, with no evidence of flame spread. However, more rigorous fire testing is required.

## Conclusions

The key conclusion from this research has found BC to be extremely robust for use as a 'shingled' façade material in outdoor environments. Whilst we are in the very early stages of understanding the material, even among the researchers who had prior experience of working with BC in other contexts, there was genuine surprise at its resilience, when used at scale, and its suitability for exposure to seasonally variable conditions. Nevertheless, the properties of BC shingles require full quantitative testing in relation to durability, impact of weathering, resistance to UV, flame spread and moisture retention. Some of the paints we applied to the material faded over ten months, and the edges of the BC shingles curled and distorted over time. These reactions require further creative and lab-based investigation. We plan to continue monitoring the material over a longer period and capturing data from lab and material testing relating to its performance. This will help us build a better profile of the material for use in façades, and for other potential applications.

Amongst the other findings of the research, we note that the production of BC using the Kombucha Method is considerably easier than anticipated. The majority of the published research to date, which has been largely lab-based and aimed at biomedical and small-scale applications, states that BC should be grown at controlled temperatures of 30 °C. We began our research following this guidance, but due to pressure on lab space we moved production to an unheated, though insulated, space. This change in context enabled us to realise that BC grown using the Kombucha Method is amenable to a range of temperatures. We have also been able to generate second and third harvests from the same growing media, thus requiring less nutrient input and physical intervention than initially considered. Production was also relatively contamination-free. This ability of the SCOBY to effectively care for itself hinders the growth of other opportunistic microorganisms. It also eliminates the need for sterilisation or autoclaving of growing containers and media, making the Kombucha Method a convenient and relatively cheap method of BC production.

Probably of most interest to industry is the development of the Biosandstone composite. It combines, as all composites do, the benefits of both materials; that is, the renewable production and biodegradable nature of BC with the robust characteristics of aggregates. We aim to enhance the circular nature of Biosandstone in future by trialling a range of industrial by-products as aggregates for the BC biomatrix. Composites offer the potential of expanding the properties, and hence applications, of BC.

From the outset of the prototype process we have sought the reaction of industry specialists, visitors, and passers-by with the BC prototype, and we have also been aware of our own interaction with it. BC is such a new and unusual building material that those who encountered the prototype were initially uneasy with its odd 'sliminess'. People remarked on its smell and wondered whether it would host potentially pathogenic life forms. But over time, feedback became progressively more positive. As the BC dried and people were able to interact with it directly and understand its green credentials, more of them expressed interest, remarking on its robustness and noticing how its surface offered a benign resting place for moths and bees. The BC lost its initial sharp vinegary smell and instead developed a sweet honey-like aroma.

The deliberate strategy to work with this relatively unknown material, at the scale of the building, and in the complex context of the external environment, inevitably revealed many challenges. But equally some of the challenges we thought we would face either did not arise or lost significance at scale. The benefit of such situated prototyping is that it quickly focuses on the most significant challenges. From here we can then begin to look at relevant emerging research from a range of scales. For example, we could use other strains of bacteria to produce higher yields at a faster rate; investigate hydrophobic modifications of BC to increase its waterproofness (Ybañez and Camacho, 2021); or utilise waste streams as sustainable nutrients (Tsouko et al., 2015). Beyond the lab, we can start to design production spaces fit for purpose, influenced by the food industry's production equipment, spaces and standards. And we can further trial other applications of BC as shingles: pre-dried, with and without fixings, in various formats and layouts, or in other building applications.

We now see bacterial cellulose as a viable building material, with much potential. Its nature and characteristics are, however, still to be fully explored. If we consider that humankind has been working with timber for over 10,000 years, then to address the challenges and opportunities that BC offers us, we must do so at a pace fitting to the climate emergency. In this context, the real challenge of bacterial cellulose lies less in the technical realm, and chiefly in the social. In other words, BC will require us to come together across disciplines and sectors: that is, DIY cultures, researchers, designers, contractors, and policy and legislation makers, to collaborate on new ways of working and living with this biomaterial.

## References


Amarasekara, A. S., Wang, D., and Grady, T. L. (2020) 'A comparison of kombucha SCOBY bacterial cellulose purification methods', *SN Applied Sciences*, 2(2), 240.

Berge, B. (2015) *The Ecology of Building Materials*, 2nd Edition, Routledge.

García, C., and Prieto, M. A. (2019) 'Bacterial cellulose as a potential bioleather substitute for the footwear industry', *Microbial Biotechnology*, 12(4), 582.

Gregory, D. A., Tripathi, L., Fricker, A. T. R., Asare, E., Orlando, I., Raghavendran, V., Roy, I. (2021) 'Bacterial cellulose: A smart biomaterial with diverse applications', *Materials Science and Engineering:* R: Reports, 145, 100623, ISSN 0927-796X.

Guo, A., Sun, Z., Sathitsuksanoh, N., Feng, H. (2020) 'A Review on the Application of Nanocellulose in Cementitious Materials', *Nanomaterials*, 10(12).

Hu, W., Chen, S., Yang, J., Li, Z., Wang, H. (2014) 'Functionalized bacterial cellulose derivatives and nanocomposites', *Carbohydrate Polymers* 101, 1043–1060

Laavanya, D., Shirkole, S., and Balasubramanian, P. (2021) 'Current challenges, applications and future perspectives of SCOBY cellulose of Kombucha fermentation', *Journal of Cleaner Production*, 295, 126454.

Portela, R., Leal, C. R., Almeida, P. L., Sobral, R. G. (2019) 'Bacterial cellulose: a versatile biopolymer for wound dressing applications', *Microbial Biotechnology,* 12(4), 586–610.

Timeanddate (1995) Available at: https://www.timeanddate.com (Accessed: 16.08.2022).

Tsouko, E., Kourmentza, C., Ladakis, D., Kopsahelis, N., Mandala, I., Papanikolaou, S., Paloukis, F., Alves, V., Koutinas, A. (2015) 'Bacterial Cellulose Production from Industrial Waste and by-Product Streams', *International Journal of Molecular Sciences.* Jul 1, 16(7), 14832–49.

Ul-Islam, M., Khan, S., Ullah, M. W., Park, J. K. (2019) 'Comparative study of plant and bacterial cellulose pellicles regenerated from dissolved states', *International Journal of Biological Macromolecules*, 137, 247–252.

Ybañez, M. G., and Camacho, D. H. (2021) 'Designing hydrophobic bacterial cellulose film composites assisted by sound waves', *Royal Society of Chemistry Advances*, 52 (11), 32873–32883.

Zhong, C. (2020) 'Industrial-Scale Production and Applications of Bacterial Cellulose', *Frontiers in Bioengineering and Biotechnology,* 8.

2.3

# 2.3

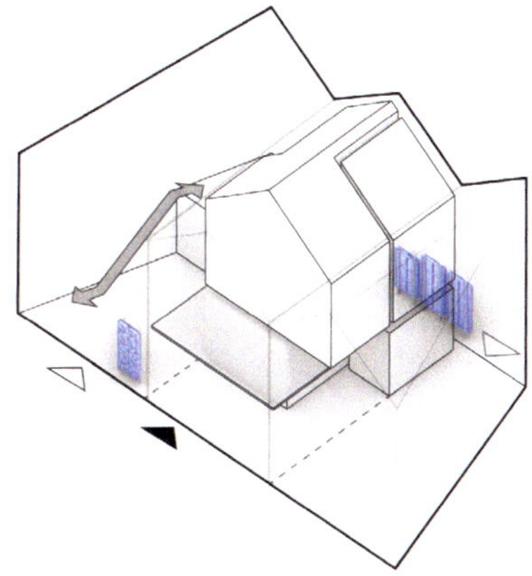

**Figure 2**
Location of the Biocellular Concrete Prototype in the OME

# Biocellular Concrete Façade
## Storing Waste and Absorbing Carbon Dioxide

**Data**
Application: Panels applied to a façade
Size of panels: 400 × 400 mm
Weight of panels: 30 kg each
Number of panels on south façade: 14
Number of panels on north façade: 4

**Team**
Ruth Morrow: Architect, Material Developer • Karolina Bloch: Architect, Design Researcher • Angela Sherry: Environmental Molecular Microbiologist • Armand Agraviador: Environmental Architecture, Computational Design • Katie Gilmour: Microbial Ecologist • Rory Doherty: Environmental Scientist, Circular Economy • Sree Nanukuttan: Civil Engineer, Concrete Technologist • Elizabeth Gilligan: Material Entrepreneur • Oliver Perry: Technical Officer

↗ **Glossary**
biomineralisation

This prototype aimed to create a 'super-green' concrete façade system. It did so by designing a concrete panel made predominantly of waste materials and by-products (90 per cent), selected for their ability to proactively support life. The concrete mix was designed to be porous, allowing water to flow through the façade, thus sustaining biological life. The panels' porosity increased the surface area and hence the potential for enhanced carbonation. Carbonation is the process by which concrete permanently sequesters or absorbs atmospheric carbon dioxide ($CO_2$), through a chemical reaction that forms carbonates. In addition, we investigated the nature of the microbial life present on the concrete and whether **biomineralisation** ↗ – the process by which living organisms produce minerals – might also contribute to atmospheric $CO_2$ sequestration.

**Figure 1**
Close-up of biocellular concrete showing texture of concrete and sedum planting

## Background – Bioreceptivity and Concrete

↗ **Glossary**
net-zero
prototype
carbonation

Achieving '**net-zero** ↗ concrete' is something of an architectural holy grail. Some of the built environment's most pressing concerns are how to decarbonise the built environment and achieve sustainable urban drainage. This **prototype** ↗ began with the conceptual aim of producing a 'super-green' concrete, made of waste streams that proactively sustain plant life. As the research progressed, the concept of concrete as a bioreceptive material developed further, to encompass hosting interdependent life forms at a variety of scales, from micro to macro. The project ran in two phases. Whilst each phase had a clear set of aims and objectives at the outset, the work was also design-led, iterative and responsive. In some regards, both the materials and the life that inhabited the concrete co-designed the process. Over time, through observation and adjustment the project has evolved, and is indeed still evolving.

**Figure 3**
Maidenhair fern and lichen growing in a wall

Bioreceptivity is defined as 'the aptitude of a material to be colonised by one or several groups of living organisms without necessarily undergoing any biodeterioration' (Guillitte, 1995). For a material to be considered bioreceptive, the conditions have to be such that living organisms can shelter, develop and multiply. To date, much of the research on bioreceptivity has been conducted under lab conditions and aimed chiefly at building and monument conservation, where the prevention of biological degradation of stone and building materials is paramount (Vázquez-Nion et al., 2018).

When intentionally designing a bioreceptive material for application in a building, a key challenge lies in the presence of water. All living organisms require water, yet this can lead to mechanical, chemical and biological degradation in buildings. Further, high levels of humidity are considered detrimental to human health, since they provide ideal conditions for the growth of fungi, viruses and bacteria. The Biocellular Concrete prototype sits squarely within the complexity that exists beyond the lab, treading a fine line between challenge and opportunity.

Of all building materials, concrete is perhaps most obviously suited to bioreceptivity, being robust, low-cost and resembling natural vertical stone habitats colonised by plants. | **Figure 3** ↗ | The apposite question is whether we should continue to work with concrete in the built environment, given its carbon-intensive profile and reliance on non-renewables. Recent research has resulted in lower-carbon concretes using cement replacements (Pulverised Fly Ash [PFA], a by-product of power stations, and Ground Granulated Blast Furnace Slag [GGBFS], a by-product of the iron and steel industries), with the possibility of using more sustainable cement replacements, such as Sewage Sludge Ash and Rice Husk Ash (Siddique and Belarbi, 2021). Indeed, the external walls of the OME are made of precast panels manufactured using GGBFS as a partial cement replacement.

Concrete absorbs atmospheric carbon dioxide across its lifetime, through a process known as **carbonation ↗** (IStructE 2020, 24–25). Present estimates suggest that during the manufacturing process, 50 per cent of all concrete's carbon emissions occur, and of that figure 2.5 per cent will be reabsorbed across its lifetime. It is also currently estimated that during demolition, concrete absorbs a further 5 per cent because of the increased surface area. The use of recycled aggregates from construction and industrial mineral waste sources leads to opportunities for local production, avoiding carbon-intensive transport. Hence concrete can become increasingly sustainable if we understand where and how carbon is created and absorbed in the cycle of manufacturing, use and end of life.

## Bioreceptive Concrete Case Studies

Bioreceptive concretes have been trialled at a range of scales. The Poikilohydric Living Walls project investigated the possibility of walls with the characteristics of 'inhabitable flesh'; that is, becoming living, visceral elements that respond to the environment and hosting life in the form of moss and algae (Cruz and Beckett, 2016). The Bio Block project investigated the use of concrete blocks as marine habitats, where, after observing the colonising species in their natural habitat, crevices of differing sizes were created in concrete blocks and left at the coastline for colonisation to occur (Firth, 2014). In more recent work, bioreceptive concrete mixes have been designed, based on readily available materials and current casting methods. Veeger et al. (2021) found that in trials, a retardant prevented the surface from curing, resulting in more porous surfaces conducive for biofilm growth (i.e. the algae growth typically found around leaky downpipes). Bone ash was also found to increase biofilm growth, and on investigation, there was little connection between the high pH of the concrete mix and its bioreceptivity.

The Biocellular Concrete prototype parallels this earlier research in considering the composition of the concrete mix and designing the surface of concrete as a habitat for life. However, it expands on this work by using proactive waste streams in the concrete mix, and pursuing an understanding of the microbial activity on the surface.

## Phase One Prototype

↗ **Glossary**
petrographic thin sections
X-ray microtomography

Phase One explored the potential of a 'super-green' concrete. It aimed to combine recycled concrete as the main aggregate with locally sourced waste streams in a manner that would sustain plant life. This meant that the concrete had to allow for water ingress and root penetration. A diverse group of waste materials was selected from a range of locally available sources. They were: egg shells, chosen as a partial cement replacement and a potential nutrient source (echoing the use of bone ash in Veeger et al.); elastic fibre waste (Lycra™), chosen as a fibre reinforcement to mitigate cracking induced by root growth; waste paint, used as a superplasticiser replacement, improving the workability of the concrete mix by allowing for an even distribution within an irregular mix of materials; recycled concrete aggregate, replacing virgin aggregates; and fly ash used as a partial cement substitute.

We tested the maximum percentage of each industrial by-product required to form a bioreceptive concrete panel. Material characterisation processes were used to understand what mechanisms were at play within the concrete and how they might affect its bioreceptivity. These involved a range of lab-based techniques, including **petrographic thin sections ↗** and **X-ray microtomography ↗**, | **Figure 4 ↗** | helping the team to understand the distribution of

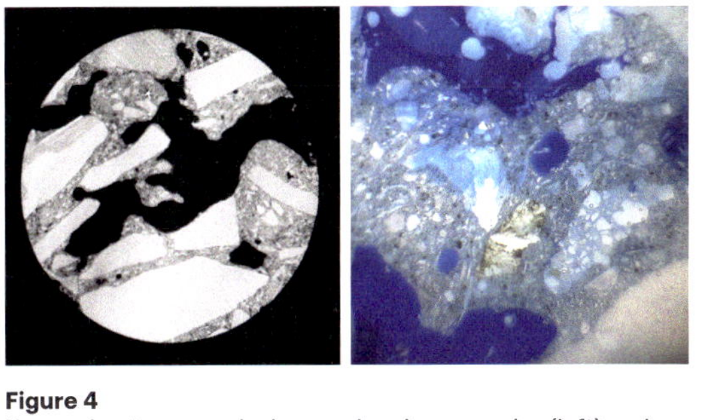

**Figure 4**
X-ray microtomography image showing pore size (left) and petrographic thin section showing the interconnected pores of the concrete (right)

materials, the pore size and the connectivity. The use of Lycra was found to result in 'shatter-resilient' concrete, ideal for withstanding the effect of plant roots, and opening up other novel applications. In the end, approximately 90 per cent of the concrete was derived from waste streams. Growth trials were then carried out, focusing on what plants could live on the concrete and how to encourage vertical growth. | **Figure 5** ↗ |

The concrete panels made up of the super-green mix were coarse in texture due to their open-pore structure | **Figure 6** ↗ | and use of recycled aggregate, and hence also thicker than standard panels. This created opportunities for recessed habitats and avoided the energy and chemical additives typically required to create the smooth, featureless concrete panels that clad many contemporary façades. In the final stage of Phase One, a pavilion was built to trial the biocellular concrete and understand growth patterns across the various microclimates of a vertical façade. | **Figure 7** ↗ |

## Learning From Phase One

↗ **Glossary**
X-ray microtomography
polymorph

The efficacy of the bioreceptive concrete, manufactured from 90 per cent waste, was demonstrated through its successful application in the pavilion and the sustained progress of plant growth over twelve months. **X-Ray Diffraction (XRD)** ↗ analysis of samples indicated that the concrete panels were absorbing carbon dioxide from the environment and producing calcium carbonate by carbonation at a higher rate than a control concrete mix made from Portland cement and recycled aggregate. Interestingly, the planted panels also contained small amounts of vaterite, a **polymorph** ↗ of calcium carbonate, produced through a biological process of carbonation, which is distinct from the carbonation caused by exposure to atmospheric carbon dioxide.

## Phase Two Prototype

The Phase Two prototype incorporated two streams of activity. The first stream involved a process of refining the design and construction of the concrete panel to better accommodate life forms, and to allow us to install panels on the façade for longer-term investigation. The second stream was focused on gaining further understanding of the bio-precipitative nature of the concrete panels and the microbial communities they sustain.

## Panel Design and Planting

The formwork and the manufacturing process of the panels were redesigned to ensure that the concrete on the front of the panels was not compacted, thus increasing the exposed surface area and potential for absorbing atmospheric carbon dioxide. In contrast, the back of the panel was designed to be compact, ensuring structural stability and allowing for ease of installation on the façade. | **Figure 8** ↗ | This was achieved through a two-pour casting process. The flat back surfaces made installation easier; however the weight of the panels, 30 kg, meant that two people were required to install them, limiting the ease with which they could be repositioned.

We designed the panel to provide habitats for a range of plant growth styles (clumping and spreading habits) and devised a hanging system that allowed us to move panels when necessary. | **Figure 9** ↗ | The panels were installed on the OME façade, with those on the south façade irrigated by a solar pump. | **Figure 10** ↗ |

A panel of industry leaders were invited to give feedback on the prototype. They stressed their interest in seeing the biocellular concrete integrated into the building structure, rather than as a bolt-on to the exterior façade. They were also interested in the application of indigenous plants rather than the sedum used from the outset. With the advice of a botanical specialist, we began an early trial with plant species found chiefly in limestone areas | **Figure 11** ↗ | and have subsequently also noted the development of algae on some surfaces of the panels. This work will continue.

Figure 5

Figure 6

Figure 7

**Figure 5**
Early-stage trial of planted concrete panels

**Figure 6**
Panels laid out in advance of hanging on the pavilion

**Figure 7**
Final pavilion from Phase 1 with control bench

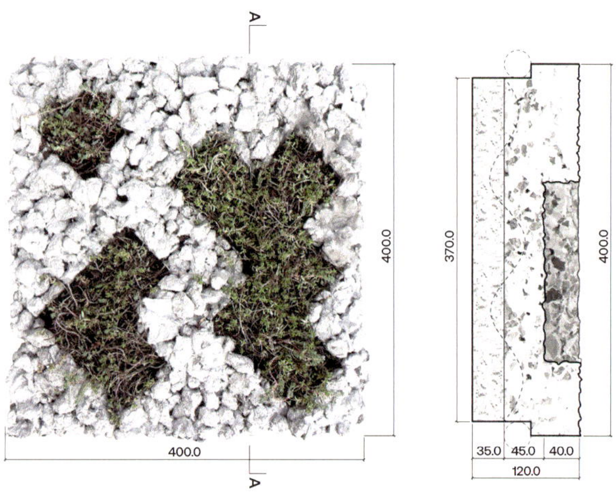


Figure 8

Figure 10

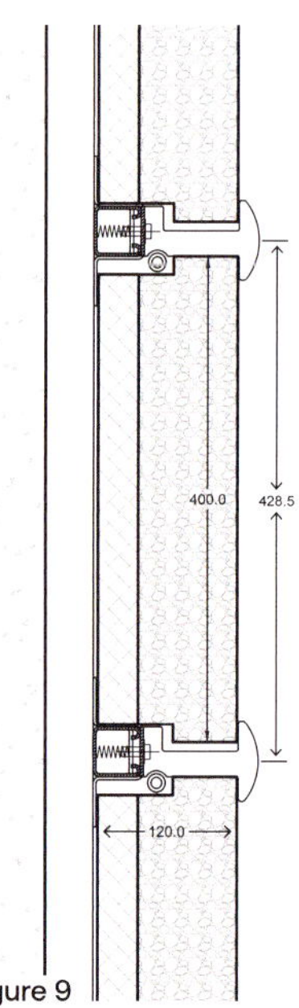


Figure 9

**Figure 8**
Design of a concrete panel consisting of rough, non compacted front layer and a compacted, structurally more stable back layer

**Figure 9**
Hanging system for the panels

**Figure 10**
Biocellular concrete panels connected via tubing and a solar pump to a large rainwater storage container

**Figure 11**
Panels planted up with moss, indigenous sedum and maidenhair fern

Figure 11

## Towards Net-zero Carbon

Phase Two of the Biocellular Concrete prototype also investigated the biologically controlled precipitation nature of the concrete panels. It did so through two stages of analysis: 1) investigating the type and nature of microbial life present, and 2) investigating the type of carbonate production.

Surface samples were taken from the concrete on the planted panels on Days 0, 103 and 197. Control samples were also collected on the same days from a non-planted panel. All samples were collected in triplicate and analysed for microbiology, XRD and stable isotope ratios.

## Understanding the Microbial Life

↗ **Glossary**
longitudinal analysis
DNA
bioinformatics

To understand the complex biological interactions associated with biocellular concrete, the microbial communities which developed in situ on the surface were investigated. Through **longitudinal analysis ↗** of the microbial communities (i.e. the same areas sampled over a period of time) we determined which communities were present on the biocellular concrete, including changes or shifts in the communities over time.

**DNA ↗** was extracted and sequenced from all samples. Subsequently **bioinformatics ↗** was used to analyse and identify the bacterial DNA sequences (Bolyen et al., 2019). Results from sequence analyses showed similarities in the microbial community composition on the concrete panels, irrespective of whether they were planted or non-planted at Days 0 and 103. However, over time the bacterial communities diverged and became dissimilar in the planted and non-planted concrete panels by Day 197. | **Figure 12 ↗** |

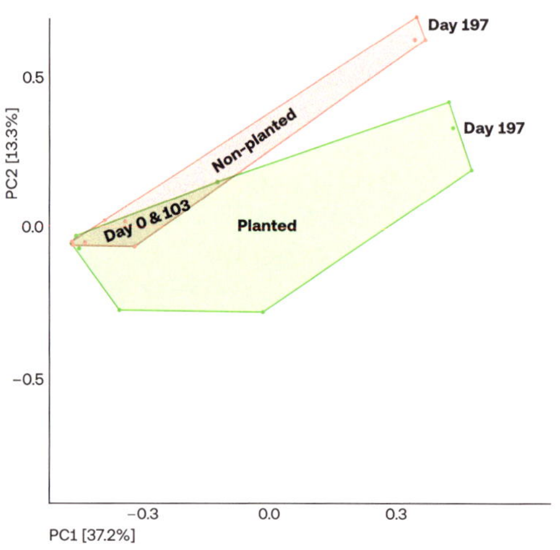


**Figure 12**
Principal Component Analysis (PCA) of the microbial community composition from bioreceptive concrete. Each marker (spot) represents the microbial community composition of a concrete surface sample, the closer the spots are to each other the more similar the microbial communities and the further the spots are from each other represents dissimilarities in the microbial community composition

At a broad level of microbial classification, the majority of concrete samples were dominated by bacteria from two major phyla, *Firmicutes and Proteobacteria.*

- Within the phylum *Firmicutes, Bacillus* were dominant on the surface of planted and non-planted panels at Days 0 and 103. Strains of Bacillus have previously been implicated in microbially induced calcite precipitation on conventional concrete (Kim et al., 2016), and more recently in self-healing concrete applications (Feng et al., 2021). Described as halo tolerant (i.e. an ability to grow in high salt concentrations) and alkali tolerant (i.e. an ability to grow at a pH greater than 9) (Chiao, Y. H., 2020), some strains are also capable of carbon dioxide sequestration (Sundaram et al., 2015).
- Microorganisms which were abundant on the surface of the planted panels after 197 days (*Paracoccus*, *Brevundimonas*, *Sphingomonas* and *Devosia* within the *Proteobacteria*) have all previously been isolated from a range of soil types, including saline and alkaline soils (Yoon et al., 2006).

## Understanding the Nature of Carbon Dioxide Absorption Using Stable Isotope Analysis

Stable isotope ratio analysis measures the ratio between isotopes of atoms with differing amounts of neutrons in their nucleus; for example the ratio between Carbon 12 and Carbon 13 atoms or between Oxygen 18, Oxygen 17 and Oxygen 16 atoms. Stable isotopes are not radioactive and do not decay, so the ratio of the isotopes acts like a signature of the chemical or biological processes that formed the compounds of interest. In the case of carbon and oxygen, these atoms make up a group of compounds called carbonates ($CO_3^{2-}$) which comprise minerals such as calcite (also known as calcium carbonate [$CaCO_3$]), found for example in limestone and concrete.

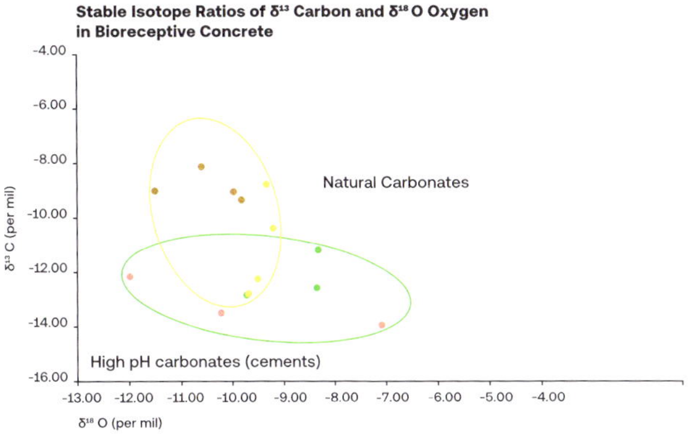


**Figure 13**
Stable Isotope Ratios of Carbon 13 and Oxygen 18 in Bioreceptive Concrete

**Key**
- Non Planted Day 0
- Non Planted Day 103
- Planted Day 0
- Planted Day 103

In | **Figure 13** ↗ |, high pH cements (carbonates) plot close to the origin of the graph and naturally occurring carbonates plot further to the upper right-hand side. As can be seen with relatively few data points to date, there is already a difference between the planted and the non-planted concrete samples. The non-planted concrete is behaving as expected, demonstrating depletion in Carbon 13 and Oxygen 18 and plotting the high pH carbonates in the bottom left of | **Figure 13** ↗ |. The planted concrete samples are becoming less depleted in Oxygen 18 and Carbon 13 compared with the non-planted concrete. This suggests that over a short time frame (approximately 100 days) there is already a biological influence on the production of the surface of the planted concrete that is producing natural carbonates. Further analysis of the same samples using X-ray diffraction also demonstrates that in the planted panel there was an increase in the amount of calcite detected, compared to a non-planted panel | **Figure 14** ↗ |. This suggests that the planted concrete enhanced the carbonation of the concrete, producing more calcium carbonate (calcite).

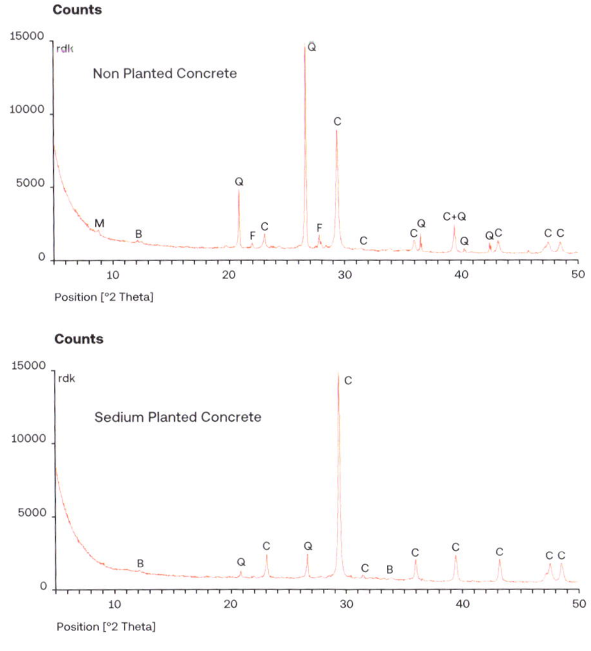


**Figure 14**
X-Ray Diffraction (XRD) analysis of planted and non-planted concrete highlighting the minerals present

**Key**
C Calcite
Q Quarz
M Muscovite
B Brownmillerite
F Feldspar

### Microbiology and Mineralogy Outcomes

Alkali- and salt-tolerant microorganisms (i.e. alkaliphilic and halophilic) were identified in association with the biocellular concrete suggesting the surface is conducive to microbial growth despite a highly alkaline pH. This is in keeping with the findings of Veeger et al. (2021). Such microorganisms are referred to as extremophiles, and their detection on the panels is as a result of the highly alkaline environment of the concrete (pH 10.2) and presumably due to salts within the concrete mix or salt release due to outdoor weathering of the panels. There is also evidence to suggest that the dominant microbes identified on the concrete surface, Bacillus, are involved in carbon dioxide sequestration and calcium carbonate precipitation in concrete environments.

**Figure 15**
View of Biocellular Concrete Panels on South Elevation

Furthermore, over the short period of this experimental work (~100 days) there is already stable isotope and mineralogical evidence of biological activity that results in the formation of calcium carbonate. The carbon and oxygen isotope ratios of the planted concrete are already beginning to diverge from those of non-planted concrete. | Figure 13 ↗ | XRD suggests that there is not only evidence of calcium carbonate due to normative concrete carbonation processes here but there is an additional mechanism of calcium carbonate formation which appears to occur only in the planted concrete.

## Conclusions

↗ **Glossary**
sustainability
microbiome
bioprecipitation

Net-zero concrete as a deliverable product still has some way to go, yet the research investigated and proposed through this prototype presents a way forward. The next steps will include better understanding of the porosity and permeability of the concrete mix to allow for both flow and retention of water, and hence life. Further experimentation is required on concrete mixes, where waste streams can contribute proactively to the creation of environments that suit diverse ecologies of plants and microorganisms. Advanced understanding of the carbonation processes that occur in concrete with high levels of waste is required, alongside the quantifiable impact of 'rougher' surfaces on $CO_2$ sequestration and the presence of water. And finally, we need to better understand the biological complexity of biocellular concrete. This will include an increased understanding of microbial communities that develop within and on the surfaces of bioreceptive materials and whether in future, specific microbes may be encouraged to enhance $CO_2$ sequestration capabilities.

The Biocellular Concrete prototype suggests key factors that could contribute towards achieving greater **sustainability** ↗, if not yet net-zero. Such a bioreceptive façade offers the potential to slow down the progress of water entering the drainage system during storms, thus reducing flooding. The microbial community within could contribute positively to the **microbiome** ↗ and to the welfare of city dwellers and, whilst more research is required, there is the potential that the **bioprecipitation** ↗ of living concrete could aid carbon capture.

## References


Bolyen, E., Rideout, J. R., Dillon, M.R., et al. (2019) 'Reproducible, interactive, scalable and extensible microbiome data science using QIIME2', *Nature Biotechnology*, 37, 852–857.

Chiao, Y. H. (2020) 'Screening for alkaline resistant spore forming bacteria as concrete healing agents', Purdue University Graduate School. Thesis.

Cruz, M., and Beckett, R. (2016) 'Bioreceptive design: a novel approach to biodigital materiality', *Architectural Research Quarterly* 20 (1), 51–64.

Feng, J., Chen, B., Sun, W., Wang, Y. (2021) 'Microbial induced calcium carbonate precipitation study using Bacillus subtilis with application to self-healing concrete preparation and characterization', *Construction and Building Materials*, 280: 22460.

Firth, L. B., Thompson, R. C., Bohn, K., et al. (2014). 'Between a rock and a hard place: Environmental and engineering considerations when designing coastal defence structures', *Coastal Engineering*, 87, 122–135.

Guillitte, O. (1995) 'Bioreceptivity: a new concept for building ecology studies', *Science of the. Total Environment*,. 167 (1–3), pp. 215–22.

Institution of Structural Engineers (IStructE). (2020). 'How to Calculate Embodied Carbon'. London. IStructE Guide https://www.istructe.org/IStructE/media/Public/Resources/istructe-how-to-calculate-embodied-carbon.pdf

Kim, H. J., Eom, H. J., Park, C., Jung, J., Shin, B., Kim, W., Chung, N., Choi, I. G., Park, W. (2016) 'Calcium carbonate precipitation by *Bacillus* and *Sporosarcina* strains isolated from concrete and analysis of the Bacterial community of concrete', *Journal of Microbiology and Biotechnology*, 26, 540–548.

Siddique, R., and Belarbi, R. (2021) *Sustainable Concrete Made with Ashes and Dust from Different Sources: Materials, Properties and Applications*. Woodhead Publishing Series in Civil and Structural Engineering.

Sundaram, S., and Thakur, I. S. (2015) 'Biosurfactant production by a $CO_2$ sequestering Bacillus sp. strain ISTS2', *Bioresource Technology*, 188, 247–250.

Vázquez-Nion, D., Silva, and B., Prieto, B. (2018) 'Influence of the properties of granitic rocks on their bioreceptivity to subaerial phototrophic biofilms', Science of the Total Environment, 610–611, 44–54.

Veeger, M., Ottelé, and M., Prieto, A. (2021) 'Making bioreceptive concrete: Formulation and testing of bioreceptive concrete mixtures', *Journal of Building Engineering*, 44, 102545.

Yoon, J-H., Kang, S-J., Lee, J-S, et al. (2006) '*Brevundimonas terrae* sp. nov., isolated from an alkaline soil in Korea', *International Journal of Systematic and Evolutionary Microbiology*, 56, 2915–9.

24

# 2.4

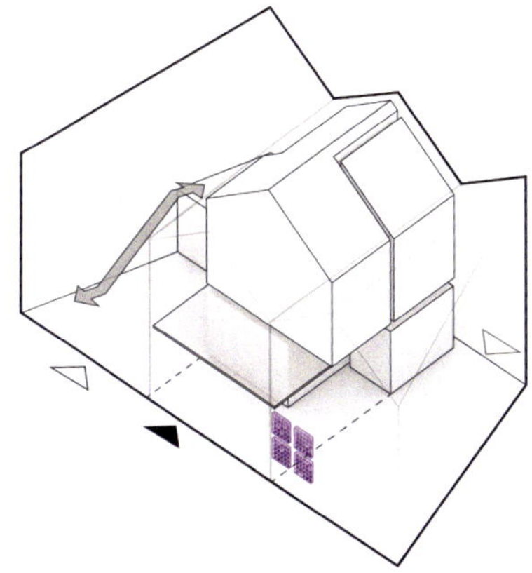

Figure 2
Location of the Healing Masonry Prototype in the OME

# Healing Masonry

## Demonstrating the Potential of Biological Self-healing for Building Conservation

**Data**
Application: Masonry
Materials: Lime mortar, pigmented dye pastes, bacteria: Sporosarcina pasteurii, sand, nutrient solution, silicone moulds, plant-based Polylactic Acid (PLA) 3D printed inserts, acrylic display
Display dimensions: 53 cm × 120 cm
Module dimensions: 5 × 5 × 5 cm cubes with a 5 mm slot on corner

**Team**
Magdalini Theodoridou: Civil Engineering, Conservation Science • Armand Agraviador: Environmental Architecture, Computational Design • Derrick Mwebaza: Engineering, Novel Applications for Vernacular Construction Materials • Angela Sherry: Environmental Molecular Microbiology, Built Environment Microbiomes • Meng Zhang: Biodesign, Microbial Molecular Biology • Martyn Dade-Robertson: Synthetic Biology, Information Architecture • Paul James: Biochemistry, Synthetic Biology

↗ **Glossary**
prototype

Figure 1
Close-up of Healing Masonry prototype

Healing masonry demonstrates the potential of biological self-healing, not only to enhance the mechanical performance and durability of masonry materials, but also to gradually transform their aesthetics. Through an exhibited installation of lime mortar samples subjected to several healing cycles, the **prototype** ↗ communicates different levels of bacterial treatment and the impact of time passing, exploring the visual communication of processes that are otherwise invisible. By bringing biological self-healing into a public context, the prototype engages a wider audience in the conservation of our built environment and heritage with emerging biotechnologies.

## A Place for Biotechnology in Heritage and Conservation

↗ **Glossary**
biomineralisation

As we become more conscious of resource scarcity and carbon footprint, the option to heal our built environment as an alternative paradigm to new construction has the potential to become a more commonplace ethical approach. The concept of healing materials and structures expands horizons for conserving our cultural heritage. According to the online ICOMOS Heritage Conservation Terminology, *conservation* includes all the processes of looking after a place in order to retain its cultural and historical significance, such as

- *to preserve*: maintain in sound condition in order to arrest decay or deterioration, or retain the components of a building that are of historical or architectural value or interest;
- *to restore*: preserve and reveal the aesthetic and historic value of the monument based on respect for original material and authentic documents;
- *to reconstruct*: reproduce, by new construction, the exact form and details of all or part of existing or vanished structures as they were at a specific period in time; and
- *to retrofit*: bring the building up to higher standards, particularly in regard to energy efficiency, security, fire protection and modern amenities.

Whether such interventions should *preserve, restore, reconstruct or retrofit* depends not only on the needs of each heritage object or site but also on the conservation strategies adopted locally. Differing cultural attitudes to impermanence, posterity and palimpsest have led to a range of contrasting treatments globally, as have varied political and economic objectives.

**Figure 3**
The Healing Masonry prototype in situ in the OME experimental building

The Healing Masonry prototype | **Figures 1–3 ↗** | investigates the application of emergent biological self-healing systems to masonry materials through a process known as Microbially Induced Calcite Precipitation (MICP), also known as **biomineralisation ↗**. Previous and ongoing research has shown evidence that MICP could be successfully used to develop self-healing masonry materials (Theodoridou and Harbottle, 2020). Rather than concealing changes over time, this prototype embraces the natural history of elements that make up the built environment, celebrating the expression of weathering and the beauty of changes from natural processes. The self-healing system being developed in the prototype repairs loss of structural integrity, thereby allowing a building to develop its own immunity. The biomineralisation process is so ingrained into the built fabric that it blurs the distinction between parts of a building that are 'alive' and parts that are inanimate. Communicating this living repair process creates an interesting duality of improving physical condition while aesthetically calling attention to a history of ongoing degradation often hidden by regeneration, and celebrating this within its built context.

## Microbially Induced Calcite Precipitation

↗ **Glossary**
nucleation
urease
hydrolysis

Biomineralisation is a naturally occurring process in living organisms that leads to the creation of biominerals. Biomineralisation is commonly divided into two categories according to the formation process: **1.** biologically controlled mineralisation in which the metabolic activity of the cells completely controls composition, localisation, **nucleation ↗** and morphology of the minerals (examples include mollusc shells), and **2.** biologically induced mineralisation where indirect precipitation of biominerals occurs, due to the interaction of metabolic by-products of microorganisms and ions present in a particular environment, such as the formation of limestone caves. The second process is Microbially Induced Calcite Precipitation (MICP). MICP through bacterial **urease ↗** is well studied in the literature. Several different reactions occur during this process. Initially bacterial urease triggers the **hydrolysis ↗** of urea in the environment to produce carbonate ions and ammonium ions. Carbonate ions then react with calcium ions to form calcium carbonate molecules which surround the bacterial cells. Accumulation of ammonium ions increases the pH and accelerates the precipitation of the calcium carbonate, forming calcite crystals. | **Figure 4 ↗** |

There are many factors that control and guide the MICP process. These include the type of microbe, calcium ion concentration, urea concentration, other chemical elements within the surrounding environment, bacterial cell number, temperature and pH.

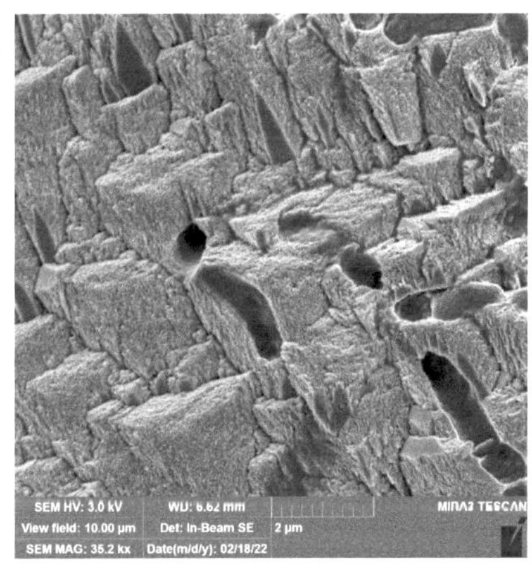


**Figure 4**
Scanning Electron Microscopy aspect of the surface of a geomaterial undergoing MICP, showing precipitated calcite crystals and imprints of bacteria

## How to Help Bacteria Thrive

The Healing Masonry prototype takes as its starting point the naturally occurring soil bacterium *Sporosarcina pasteurii*, a well-known mineralising bacterium frequently associated with MICP. *Sporosarcina* has the ability to induce calcite precipitation through the process of biological cementation of solids, when provided with a calcium source and urea | **Figure 5 ↗** |. Commercially available laboratory reagents include calcium chloride, calcium oxide, calcium acetate, calcium nitrate and urea, although environmental wastes also offer alternative sources; for example, cow urine as a form of urea (Casas et al., 2022). *Sporosarcina* has been identified as instrumental in the MICP process for biocementation since its initial growth conditions were published (Gibson, 1934). However, research into defining the correct balance of nutritional requirements for optimal biomineralisation is still ongoing in an attempt to ensure the economic feasibility of this biotechnology (Lapierre et al., 2020). The concentration of nutrients in healing cycle solutions, specifically calcium chloride ($CaCl_2$) and urea, can have a profound effect on *Sporosarcina* growth and metabolism, and consequently the MICP process (Ma et al., 2020, Lapierre et al., 2020). It has been suggested that higher concentrations of urea-CaCl2 solutions promote a more efficient MICP process through calcite precipitation, filling all gaps and spaces within a structure, thereby providing strength (Gebru et al., 2021). Biomineralisation by *Sporosarcina pasteurii* in the presence and absence of urea has also been assessed (Ma et al., 2020). Cells grown in an alkaline environment with urea showed faster growth, a more robust cell structure and increased negative surface charge, promoting calcium ion binding. Calcium ions are an important component in biomineralisation. When microorganisms secrete metabolic products into their surrounding environment, the products react with the calcium ions resulting in the precipitation of minerals.

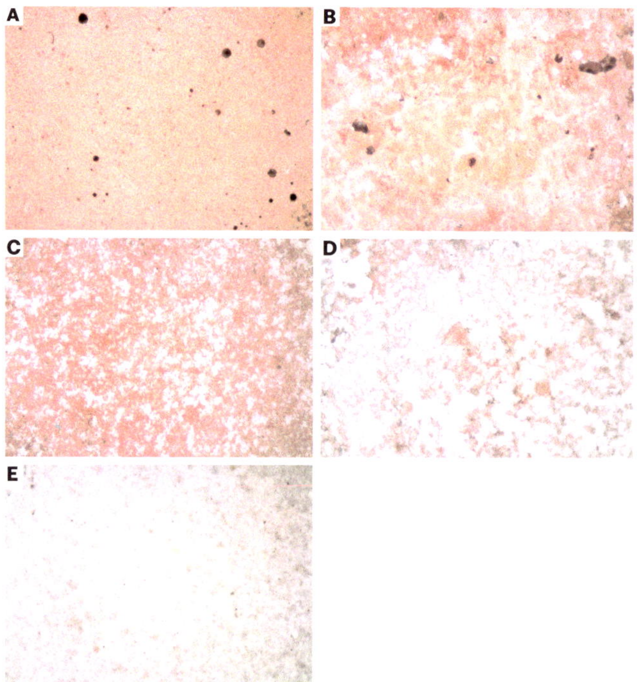


**Figure 5**
Microscope images of changes in surface texture of lime mortar over time when the resident bacteria were fed nutrients over several cycles. Surface texture prior to treatment (a) and after 7, 14, 21, and 28 days of treatment (b to e, respectively)

## Designing for Aesthetics and Experimentation

Through an exhibited installation of lime mortar samples treated with *Sporosarcina pasteurii* to induce MICP, this research sought to engage a wider audience in the repair of the built environment, specifically historic environments. It aimed to go beyond demonstrating mechanical performance and durability and explore the potential of visual communication of processes that are otherwise invisible. The prototype is thus designed to promote discussion around:

- Legibility: exploring and conveying the gradual effect that a microbiological process has on an inorganic construction material;
- Material performance: demonstrating improvement on, not obsolescence of, previous technologies, and highlighting the resilient potential of making a conventional material smarter through biotechnology;
- Historic sensitivity: discussing where implementation fits into discourse around intervention, conservation and built heritage; and
- Applicability: studying systems when applied in real world conditions, and natural climatic and biological processes in different settings.

The design of the prototype visually communicated temporality and variability by representing different levels of bacterial treatment and the impact of time passing.

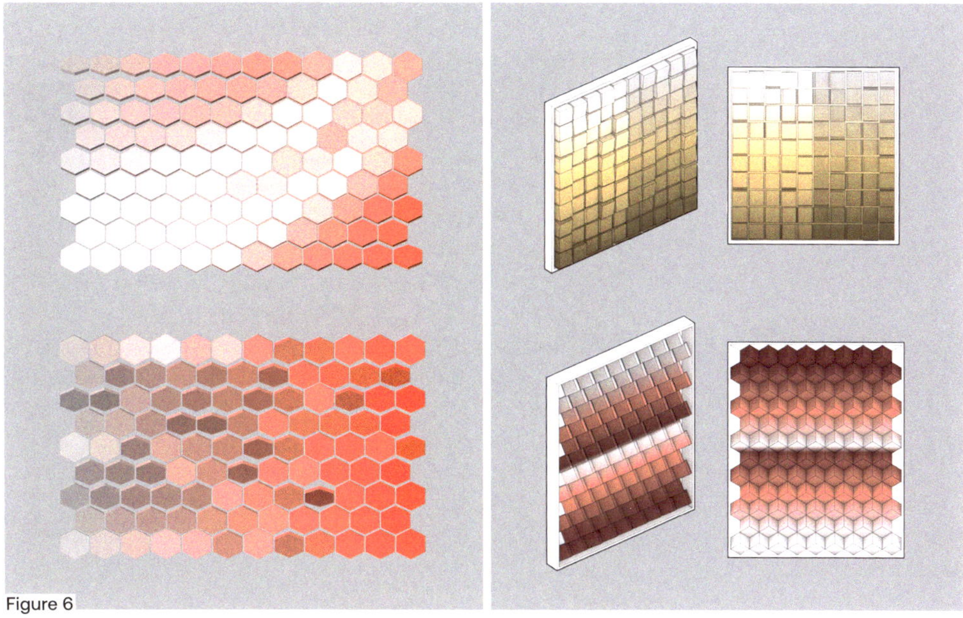

Figure 6

Figure 7

**Figure 6**
Initial hexagon designs for the prototype

**Figure 7**
Visualisations of display approaches

Initial designs for the prototype took the form of tiles | Figure 6 ↗ |. Their display method explored an intersection of regimented distribution and playful articulation, balancing the scientific method with the chaotic reality of a real-world application. However, it was thought that the patterned angular articulation of the tiles might distract from the controlled gradation of colour.

This evolved into cube modules, providing a more standardised unit on which to later perform physical tests, such as strength and durability. Similar to the tile design, a regimented but articulated approach was explored and in addition each unit was designed to be demountable for inspection. The final design took demounting and inspection further in the form of acrylic pigeonholes that celebrated each sample individually. Corner-fixed units allowed rotation for an observer to have a clear view of every facet and remove and replace each sample securely while the acrylic allowed a more uniform light to better highlight surface textures and colour variation. | Figure 7 ↗ |

### Preparation of Lime Mortar Cubes

Lime mortar was prepared by mixing natural hydraulic lime with fine aggregates (standardised silica sand) at a ratio of 1:3 in a planetary mixer for 180 seconds | Figure 8 ↗ |. Water was added to maintain the mortar at a workable consistency at a flow rate of 165 ± 5 mm (according to BS EN 1015-3:1999). Various pigments were added to the mortar mixtures | Figure 8 ↗ |; however, a commercially available toning powder (red) was eventually selected for this study because it produced a consistent colour of units and contrasted with the colour of calcium carbonate precipitation (white) | Figure 9 ↗ |. Lime-based mortar mixed with the pigment was cast into cubic moulds (50 × 50 × 50 mm) as seen in Figure 8, resulting in solid, identically sized units after curing. Upon casting, the units were compacted by tilting the moulds through approximately 30°, then tapped ten times, returned to the horizontal, and then tilted and tapped a further ten times. Half of the units were imprinted with a surface texture on at least one of the cube faces during casting or whilst solidifying | Figure 9 ↗ |, after which the units were cured by covering them with a damp hessian cloth and placed in a sealed humid box for 28 days.

### Surface Textures

**↗ Glossary**
Voronoi pattern
Shortest walk branching pattern
Reaction-diffusion (Turing) pattern
Phyllotaxis pattern

The prototype aimed to observe the effect of calcite precipitation on both visible fissures on the surface of the lime mortar and microcracking on perceptibly uniform surfaces | Figure 10 ↗ |. To investigate this, half of the cubes featured an articulation pattern imprinted into the surface either through a mould insert or through embossing while the mortar was solidifying, but still soft. Generated using visual programming, the mould inserts were 3D printed using coloured (plant-based) Polylactid Acid (PLA) filament and designed to represent varying intricacy and fissure depths and widths, through different propagation motifs taken from nature. Examples included:

1. **Voronoi pattern ↗**: a tessellation of regions where all points within each region are closest to a randomly scattered reference point – found in fissure propagation in ground shrinkage;
2. **Shortest walk branching pattern ↗**: a generative path connecting closest defined spatial nodes, which split at given intervals as the path approaches a destination point – found in tree branches and roots;
3. **Reaction-diffusion (Turing) pattern ↗**: spatially periodic patterns that arise from random or uniformly homogeneous distributions of two diffusible domains interacting with each other – found in chromatic patterns in animal skins; and
4. **Phyllotaxis pattern ↗**: spiralling patterns that arise from intersecting regular intervals in polar coordinate geometry – found in botanical structures such as pine cones and aloes.

Figure 8

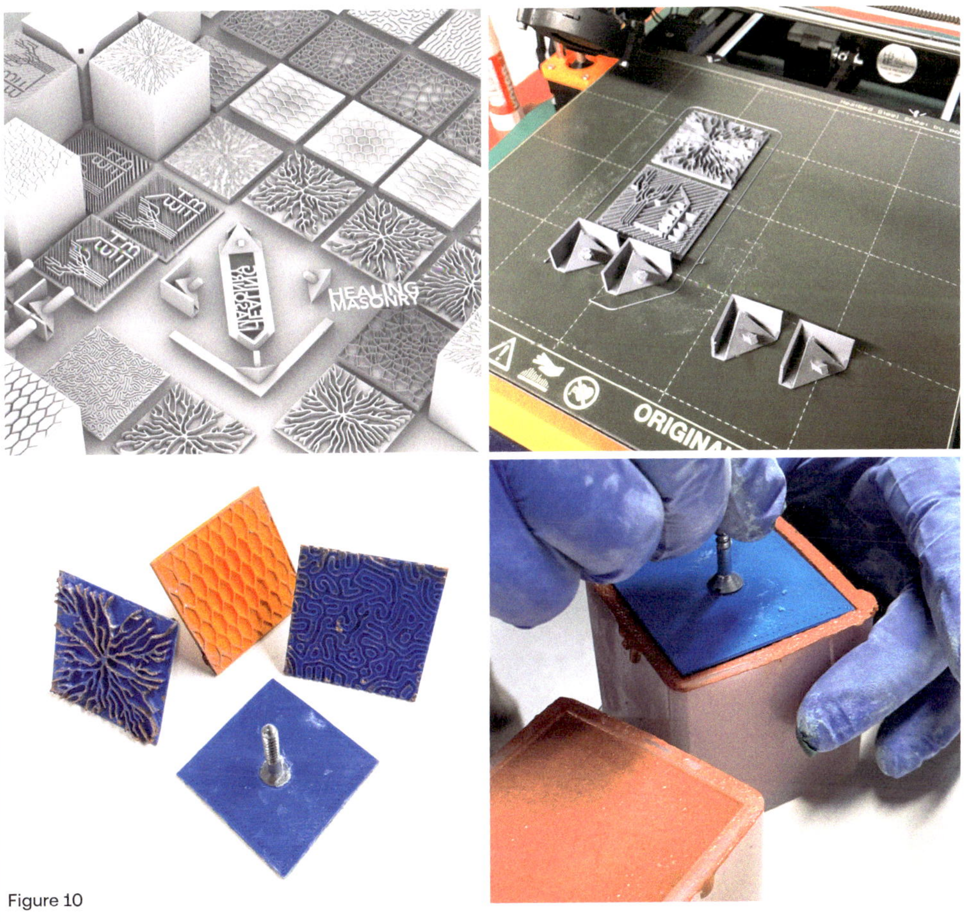


Figure 10

**Figure 8**
Production of lime mortar cubes

**Figure 10**
A series of 3D printed inserts were designed to imprint onto half of the cubes to observe calcite precipitation around varying depths and widths of fissures

**Figure 9**
A range of lime mortar cubes were created whilst optimising and testing conditions of consistency, texture application and pigmentation prior to being used to study calcite precipitation

## The Biological Healing Process

Following curing, the units were subjected to biological healing cycles, comprising bathing the units in a bacterial treatment solution, containing the bacterium *Sporosarcina pasteurii*, followed by the nutrient solution containing calcium chloride, which are both important for the efficiency of the healing process. Given the conducive conditions, microorganisms can alter their environment to induce the production of calcium carbonate. As mentioned earlier, this process is known as Microbially Induced Carbonate Precipitation (MICP). The units were then removed from the solution and dried overnight in an incubator (30 °C). After this, they were immersed daily in the nutrient solution containing calcium chloride for 28 days, which was refreshed every seven days. For each biological healing cycle, the units were stored overnight in an incubator (30 °C).

Images of the units were taken on Day 0 prior to immersion in solutions and subsequently monitored daily to record visual modifications on the surface of the units. After every seven days, six units were removed and discontinued from further treatment to evaluate the effect of the healing cycles on the surface of the units at 7, 14, 21 and 28 days.

## Observations From Variables

↗ **Glossary**
deposition

Lime mortar is a porous material, and as such it has the potential to absorb a substantial amount of both the bacteria and nutrients required for biological healing. Upon immersion in the bacterial solution, the units exhibited a high absorption potential without damaging them. When the units were immersed in the nutrient solution containing calcium chloride, a whitish powder was immediately formed on the surface of the unit but was then resuspended into the solution.

Prior to biological healing (Day 0) and for two to three days following biological healing, no visibly noticeable calcite formed on the surface of the units. However, after seven days, a stable deposit of calcite was observed on the surface of the units. Following further biological healing cycles, there was a visible sequential increase in calcite precipitation on the surface of the units over time, with cubes immersed for 28 days showing more calcite deposited on their surfaces compared to those treated for 7, 14 and 21 days | **Figure 5** ↗ |. A gradual colour change from red to white was observed following successive biological healing cycles | **Figure 11** ↗ |.

It was further observed that for units with surface textures, calcite **deposition** ↗ initially occurred in the fissures compared to raised sections of the texture on the units | **Figure 12** ↗ |. This could be attributed to the ability of the fissures to embrace and store more bacterial cells and nutrients in a smaller surface area; therefore precipitation rates presumably increased. This indicates the potential of biomineralisation to be employed for the healing of damage in the built environment, which not only helps to protect the materials from further deterioration, but could also preserve their marks in case such damage is of historical significance.

For display in the OME, the cube units were arranged randomly | **Figure 13** ↗ |. The playful idiosyncrasy of masonry construction was still referenced in the unique rotation of each cube.

Figure 11

Figure 12

**Figure 11**
Changes in calcite deposition over time on the surface of lime mortar cubes (top row) and imprinted/textured lime mortar cubes (bottom row). From left to right, cubes display calcite deposition after 7, 14, 21 and 28 days of treatment cycles, respectively

**Figure 12**
Calcite deposition was initially concentrated in the crevices of the lime mortar surface textures

**Figure 13**
Healing Masonry prototype on public display, demonstrating variation in the visual appearance of the lime mortar units following a series of biological healing cycles

## Future Directions – Engineered Bacteria for Improved MICP and Self-healing

One of the biggest challenges to the use of MICP in the built environment is improving process efficiency. While researchers continue to investigate this, approaches have not yet resulted in a solution that rivals traditional cement production. For current industry models to accept MICP and biological self-healing, the speed of the biological processes will need to be improved, the cost reduced, and polluting by-products kept to a minimum or eliminated.

Most research to date has focused on recreating the *Sporosarcina* urease system in traditional laboratory strains of bacteria (*Escherichia coli or Bacillus subtilis*), with limited success. Recent advances in engineering biology could allow the development of genetic toolkits that enhance MICP production in *Sporosarcina*. By augmenting *Sporosarcina* MICP-producing systems, we could design and engineer the bacteria to be more efficient, to grow on waste materials, to improve $CO_2$ sequestering processes and to reduce the production of harmful by-products (e.g. ammonia). Further, it may be possible to produce materials that have enhanced physical properties, such as strength, durability, and pigmented healing, and to tailor the self-healing properties of the material to respond to different stimuli.

## A New Paradigm in Expressing Immunity in Buildings

Through the Healing Masonry prototype, enhanced biotechnological systems are being developed and, for the first time, not only evaluated for their physico-mechanical performance but also 1) highlighted for their unique function at a larger scale, 2) exploited for their interaction with the visual and tactile expression of the materials, and 3) taken outside the lab to understand their coexistence with other biological processes.

With the lime mortar units now on public display | Figure 13 ↗ |, the public can view and appreciate biomineralisation and self-healing in typical settings of the built environment. Although widely researched in other applications, MICP has always been a relatively specialist and non-visible process. Yet it has the potential to improve the strength and durability of various construction materials. The Healing Masonry prototype brings this biotechnology to the forefront allowing it to be understood and appreciated by the public and to visibly demonstrate the potential of microbiological processes and self-healing to building professionals, policy makers, end users and the public more generally. This is witnessed in the considerable interest that the Healing Masonry prototype has already generated from non-specialists and building professionals alike.

Rather than concealing changes over time, the Healing Masonry prototype embraces the natural history of the materials that make up the built environment. Through a visible biological record, Healing Masonry enhances and maintains our cultural heritage by calling attention to the beauty of changes brought about by deterioration and immunity in buildings.

## References


British Standards Institution (1999) 'BS EN 1015–3:1999 'Methods of test for mortar for masonry Determination of consistence of fresh mortar (by flow table)'. London: BSI.

Comadran-Casas, C., Schaschke, C. J., Akunna, J. C., Jorat, M. E. (2022) 'Cow urine as a source of nutrients for Microbial-Induced Calcite Precipitation in sandy soil', *Journal of Environmental Management*, 304, 114307.

Gebru, K.A., Kidanemariam, T.G., and Gebretinsae, H.K. (2021) 'Bio-cement production using microbially induced calcite precipitation (MICP) method: A review', *Chemical Engineering Science*, 238:116610.

Gibson, T. (1934) 'An Investigation of the Bacillus Pasteuri Group' *Journal of Bacteriology*, 28, 491–502.

Lapierre, F. M., Schmid, J., Ederer, B. Ihling, N., Büchs, J., Huber, R. (2022) 'Revealing nutritional requirements of MICP-relevant Sporosarcina pasteurii DSM33 for growth improvement in chemically defined and complex media', *Scientific Reports*, 12(1):2017.

Ma, L., Pang, A. P., Luo, Y., Lu, X., Lin, F. (2020) 'Beneficial factors for biomineralization by ureolytic bacterium *Sporosarcina pasteurii*', *Microbial Cell Factories*, 19:12.

Theodoridou, M. and Harbottle, M. (2020) *Self-healing geological construction materials and structures.* Cordis EU Research Results, European Commission, Project ID 745891. Available at: https://cordis.europa.eu/project/id/745891 (Accessed: 29 September 2022).

23

# 2.5

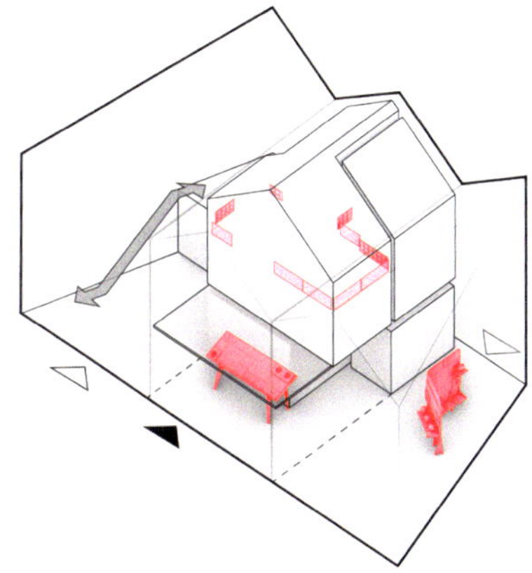

**Figure 1**
Location of BioMateriOME in the OME experimental living house

# BioMateriOME
## Monitoring and Perception of Microbe-material Interactions

**Location**
BioMateriOME is split into two aspects, Public (BMP) and Experimental (BMX). The Public aspect consists of a variety of modular display elements spaced around key open areas of the OME. The modules are moveable and can be attached together. The Experimental aspect is limited to the studio apartment and bathroom, and located in strategic places of specific activity.

**Data**
Application: A tactile material library twinned with microbiological monitoring of surface materials.
Materials: plywood displays, metal fixings, paper, samples
Display dimensions: BMP 1 m × 1.7 m, BMX 1 m × 0.4 m
Module dimensions: BMP 20 cm, 25 cm and 50 cm squares

**Team**
Angela Sherry: Environmental Molecular Microbiologist, Indoor Microbiomes ● Beatriz Delgado-Corrales: Environmental Molecular Microbiologist, Indoor Microbiomes ● Romy Kaiser: Material Designer, Biomaterials ● Paula Nerlich: Material Designer, Circular Biomaterials ● Armand Agraviador: Environmental Architect, Computational Design ● Oliver Perry: Technical Officer

**↗ Glossary**
prototype
microbiomes

The BioMateriOME **prototype**↗ has been co-designed to shed light on the microbial world with which we share our indoor living spaces by providing accessible solutions that increase societal awareness and understanding of indoor **microbiomes**↗. BioMateriOME combines the elements of a public materials library with microbiological monitoring of surfaces to characterise microbes inhabiting conventional or novel surface materials and, in parallel, to gauge human perceptions of microbe-material interactions in the built environment. Combined outputs from BioMateriOME will contribute to responsible practices in design and construction which could eventually be codified into benchmark standards and building regulation.

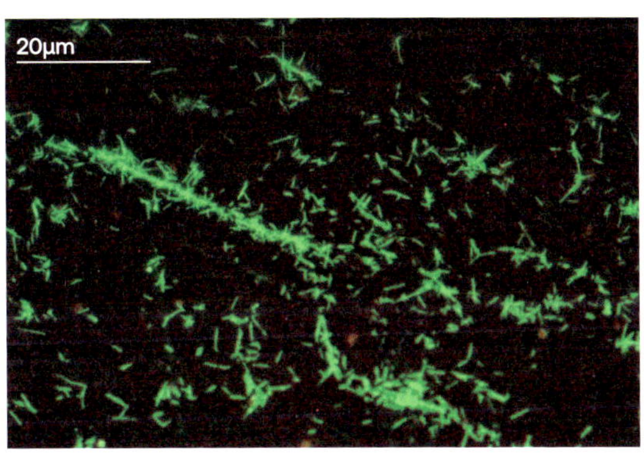


**Figure 2**
Closeup of microbes on a glass surface

## Introduction – Our Unseen Cohabitants

↗ **Glossary**
ecosystems
biofabrication
living construction

In a world adapting to the aftermath of a global viral pandemic, what do we really know about the microbes that share our homes, indoor spaces and surfaces? The survival of microbial cells has previously been investigated on a range of household surfaces including glass, stainless steel, cotton, polyester, wood, vinyl, plastic, and ceramics (Coughenour et al., 2011; Gibbons et al., 2015; Traoré et al., 2002) | **Figure 2** ↗ | and recently the persistence of Coronaviruses on inanimate surfaces has received much attention (Fiorillo et al., 2020; Kampf et al., 2020). Research has shown that those growing up or living in rural (outdoor) environments have a decreased risk of allergies due to exposure to a more diverse environmental microbiota, compared to those living in densely built urbanised environments (Kirjavainen et al., 2019; Parajuli et al., 2018). Furthermore, changes in building design, hygiene practices and urban life have recently been recommended as ways to alter the microbiomes of our homes, with the authors suggesting that 'the microscopic elements of "indoor **ecosystems** ↗", and how they are created and maintained, must become a focal point for research' (Wakefield-Rann et al., 2020).

Considering this, BioMateriOME evolved from discussions between an interdisciplinary team of material designers, architects and molecular microbiologists. BioMateriOME has two aspects, located in the OME:

- BioMateriOME-Public (BMP) – a *multisensory materials library* which is a public-facing installation, and
- BioMateriOME-Experimental (BMX) – a *panel of material installations* designed to perform experimental studies into the development of surface microbiomes in the home.

We aim to co-create research solutions to address the societal challenges of understanding and comprehending indoor surface microbiomes in the built environment | **Figure 3** ↗ |.

BioMateriOME seeks to use engagement with the public to promote a more nuanced view of microbes in human habitats and remove their stigma by nurturing curiosity and material intelligence and supporting positive behaviour towards microorganisms. By exhibiting materials in a multisensory way (tangible, visual and digital), the Public materials library (BMX; | **Figure 4** ↗ |) will break down barriers of knowledge and relink aesthetic and technological goals which have been historically separated into arts and science (Miodownik, 2015). The Experimental aspect (BMX; | **Figure 4** ↗ |) aims to shed light on the invisible microbial world which exists on common indoor surface materials. We also investigate innovative construction materials, including biocomposites, textiles and **biofabrication** ↗ materials, which are transformative and are gaining traction as '**living construction** ↗' materials (Dade-Robertson, 2021).

BioMateriOME introduces the concept of microbially aware design in architecture, acknowledging that, as human inhabitants, we do not live in isolation but play a role in an ecosystem, albeit invisible in our habitats. BioMateriOME seeks to open a discussion on how living materials and inert materials that harbour life can incorporate dimensions of human perception, ergonomics, legibility, and wellbeing in an accessible as well as rigorous way through Public and Experimental aspects of the project. Microbially aware design seeks to address the following questions:

- How can we provide an added scientific dimension to current cultural practices in designing spatial adjacencies, placement of utilities and circulation (both human and mechanical, electrical, plumbing) in architecture?
- Should we give special consideration to tactility and sensory relationships within our built environment – what about small-scale design?
- In what new ways can health and wellbeing influence architectural design to accommodate projections for future technologies, behaviours and pandemics? And could this be codified into benchmarks and regulations?

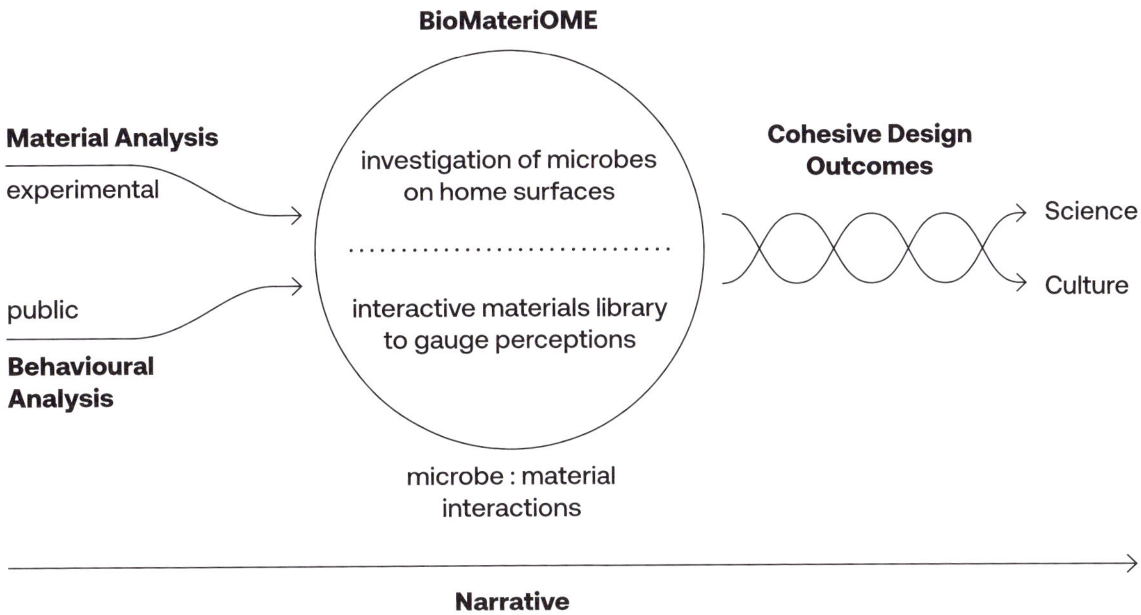


Figure 3

Figure 4

**Figure 3**
The concept of BioMateriOME. Microbial data generated from the surface of materials in the experimental aspect of the prototype will be digitally-twinned with materials in the public library in order to make microbiological data more accessible to society

**Figure 4**
Elements of the BioMateriOME prototype in situ in the _OME experimental building (BMP up, BMX down)

## Design and Curation of the Public Materials Library (BMP)

↗ **Glossary**
hygromorphs

The design of BMP responds to a set of conceptual and practical parameters based on current developments in material design, material trends and possible future visions (Dade-Robertson, 2021; Solanki, 2020; Franklin and Till, 2018). In contrast to traditional library sorting principles, a cluster system was devised during the development of BMP which organises materials according to both the chronology of their development and their state of 'being alive' (Trubiano, 2006). Clusters of 'Speculative' versus 'Applied' and 'The Living' versus 'The Lived' are surrounded by time-related categories of Future Vision, Future System, Trend Material and Experience | **Figure 5** ↗ |. Organising the materials around these concepts serves a dual purpose by 1) providing an overview to understanding developing material trends, and 2) setting guidelines for the curation of the public materials in the library.

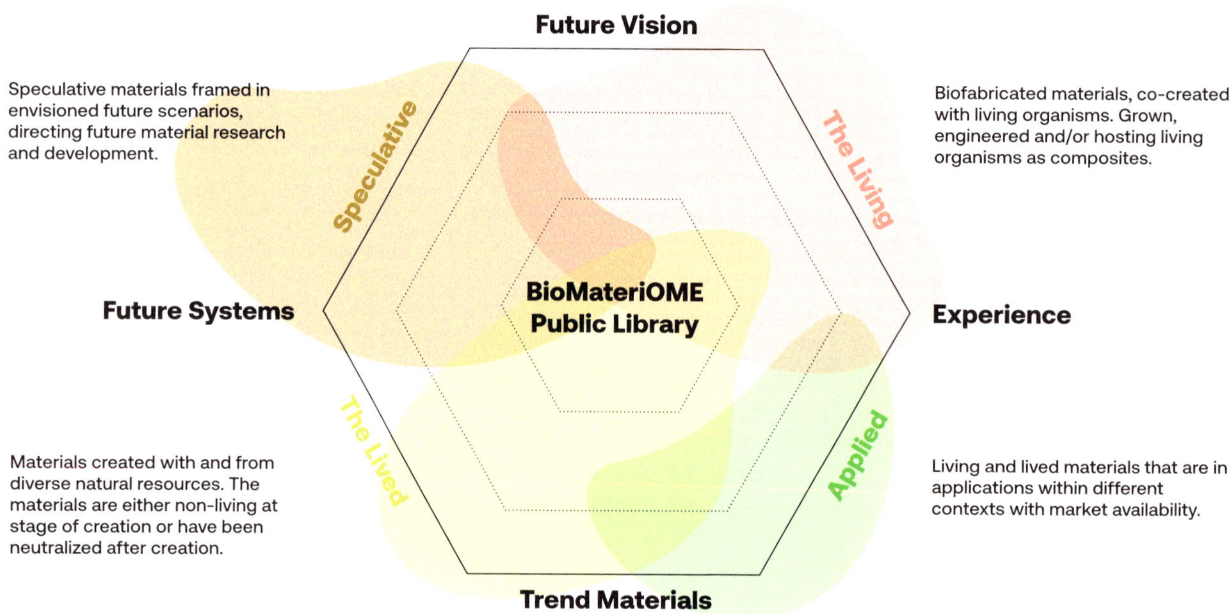


**Figure 5**
Classification of material characteristics for curation of the public library and to understand developing material trends

'The Living' comprises biofabricated materials that are co-created with living organisms during the construction process, or can even be considered as 'alive' during the time of use. This includes grown, engineered and hybrid living materials or composites with a lifelike appearance, which are able to be animated or contain live components such as photosynthetic biocomposite ceramics (Crawford et al., 2022). Responsive and active materials are therefore clustered within this category. These are materials which adapt to their environment in a seemingly living manner, for example **hygromorphs** ↗, which react to humidity by changing shape. By contrast, materials in 'The Lived' cluster contain non-living materials created with and from diverse natural resources. These are materials processed towards a specific function or shape after their lifespan is over; for example, wood or cotton. The 'Applied' cluster focuses on commercially available novel or conventional biomaterials which can be implemented on an industrial scale. The 'Speculative' cluster is framed around envisioning future scenarios and is aligned with Future Systems; that is, materials which are forthcoming and may exist in commercial markets in the future as developments in (bio)technology facilitate materials research progress. The 'Experience' cluster is based on knowledge of materials which are well known and currently commercially available, which overlaps with 'The Living', which incorporates biofabricated materials co-created with living microorganisms. Current materials used often within the built environment are represented in Trend Materials, with experimental materials denoted in Future Vision | **Figure 5** ↗ |.

## Engaging with Designers, Crafters and Manufacturers

↗ **Glossary**
biophilia

BMP acts as a tool to foster discourse on sustainable materials and a future built environment through material interaction and information sharing. The intention of BMP is to gauge opinions and perspectives through feedback from the public, visitors and researchers in ways that could help to shape new materials, biotechnologies, narratives and visions that are essential for research into sustainable living materials for the future. As such, the materials library offers materials that are relevant to the UN Sustainable Development Goals (UN, 2015), as well as those that are interesting in terms of human-microbial interactions, **biophilia** ↗/biophobia and aesthetic properties. Materials are sourced from researchers within the Hub for Biotechnology in the Built Environment (HBBE), as well as external material designers, artists, researchers and manufacturers. The range of presented materials varies from soft to stiff, lightweight to heavy, opaque to translucent, sleek to textured and beyond. They are produced from diverse base materials and a number are co-created with different microorganisms. This wide spectrum invites visitors to interact with and explore current and future material possibilities.

## Material Narratives

Each material's narrative was considered in the compiling of the material library, and will continue to be expanded in the prototype by linking the tactile display to the scientific microbiological data generated, allowing participants to repeatedly assess the mapping of the library. Narrative refers to the implicit qualities that provide the backdrop of analysis. It informs why any material is chosen for any purpose. Narrative encompasses a material's origin and refinement, its life cycle, transportation and associated industries, and its sustainability. It also encompasses a material's cultural history, discovery, societal impact, ethics and future.

Mounting the materials sourced from suppliers and deciding the best way of displaying them involved a sensitivity to the relationship between the digital and physical elements of the installation. We wanted the display to foster intimate relationships with each material – both a tactile relationship and a relationship with the materials' suppliers. We also considered shipping logistics and the way each material would be archived. All of this contributed to the narrative of each material.

## Mapping Materials

In order to provide information to support each of the materials, a questionnaire was developed to gather data from the material suppliers. Utilising Sustainable Design and Material Pathways Cards in the 'design for sustainability' system (Hasling et al., 2020), the questionnaire sought information on method, material narrative, maintenance (care) instructions, as well as end of life treatment. Materials in the library are mapped against technical data for strength and resistance as well as health factors such as impacts on air quality, noise, and microbiome development. Materials are classified through a simplified scoring system which assigns core attributes, allowing an easy comparison between materials | Figure 6 ↗ |. Based on mapping parameters, material information sheets and digital signatures (quick response [QR] codes) are available for each material, displayed alongside or on the reverse of materials. The information shows ongoing material developments from the HBBE and other materials suppliers. Collecting and communicating material information through the selected mapping criteria frames the material qualities around factors that aim to positively impact a healthy built environment and sustainable, circular practices of material fabrication (Browning and Ryan, 2020).

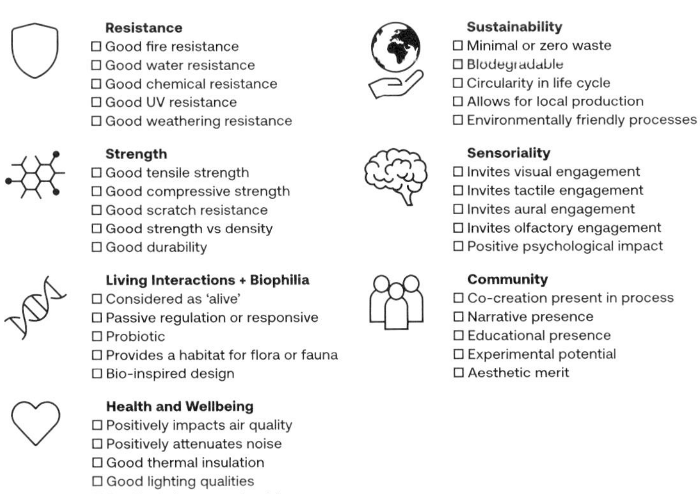


**Figure 6**
A simplified scoring system assigns core attributes that allow for easy comparison of materials

## Building BioMateriOME – Design for Engagement

↗ **Glossary**
CNC
CAD

In designing BMP, the public installation, several approaches were explored for presenting materials in a way that would work as both a curated multisensorial display and a tactile interface | **Figure 7 ↗** |. Detachable modular elements within an extendable framework were designed for this purpose. Components were fabricated by **CNC ↗** milling using **CAD ↗**, before being assembled into modules that could support both flat material samples and 3D objects | **Figure 8 ↗** |. The final design presents variations on a perforated plywood sheet into which modules can be slotted or affixed in a grid. There is an inclined desk and an upright display variation that can be independently situated or attached together with hinges to be freestanding | **Figure 9 ↗** |. The modular system invites a playful approach to the materials and opens the display layout to multiple arrangements. Our intention is for different audiences to interpret relationships between the materials mounted on the modules, with the framework acting as a blank canvas | **Figure 10 ↗** |.

The Experimental installations (BMX) are designed to echo the public elements, comprising a systematic grid of repeated materials with different surfaces, composed of conventional home surfaces as well as novel biomaterials; for example, mycelium-covered tiles | **Figure 11 ↗** |. These panels require an additional level of control which is achieved by mounting the materials high enough that they are out of reach in a domestic setting (in this case, at the height of a picture rail within the OME experimental studio, | **Figure 4 ↗** |). The dimensions for these panels make the most of the remaining plywood, following the milling of the public modules. Most panels are mounted around the perimeter of the experimental studio, with some panels placed in locations of particular interest to monitor microbiome development; for example, in the bathroom.

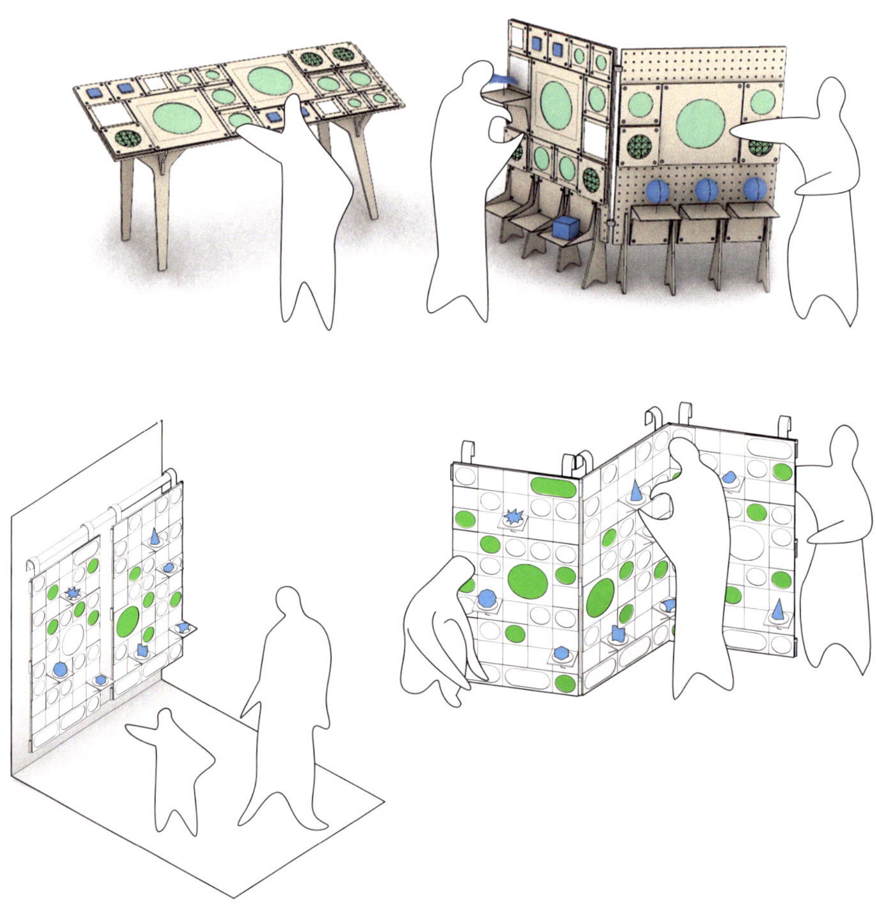

Figure 7

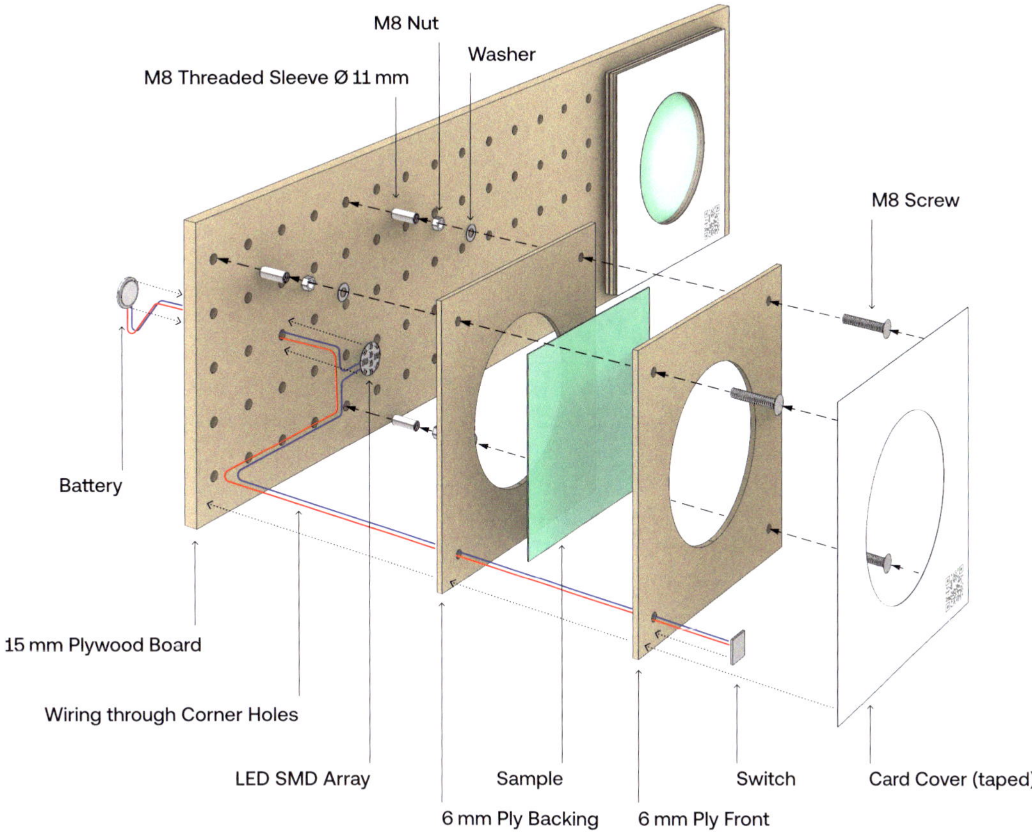


Figure 8

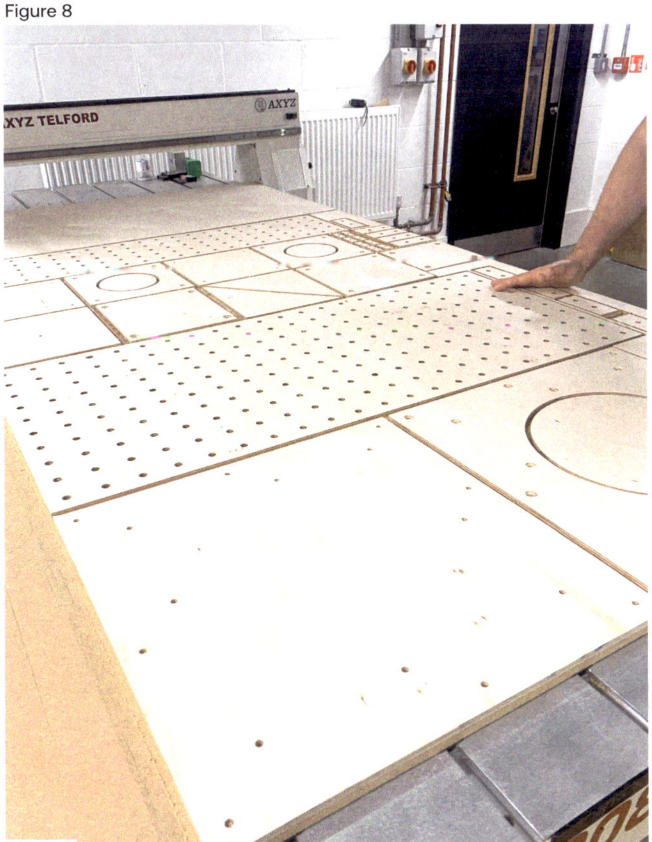


Figure 8

**Figure 7**
Several approaches were explored to express the material library in a way that invites interaction while making information available

**Figure 8**
Care was taken to provide the best platform for the material samples to be both celebrated and accessible. Assembly was kept low tech and modular for greater flexibility as the library grows

Figure 9

Figure 10

Figure 11

**Figure 9**
The Public library (BMP) displays the visual as well as multi-sensory qualities of many materials. Further information was available for all materials via digital quick response (QR) codes

**Figure 10**
The removable modules have material classification information on the reverse. Variants of the modules have lighting and multimedia options

**Figure 11**
The Experimental installations were placed at strategic locations and at height within the experimental studio apartment in the OME, where microbes will settle and develop over time

## Potential Technological and Cultural Futures

↗ **Glossary**
probiotic

Our habitats are shaped by vernacular and industrialised design approaches that have their own implicit conventions and explicit rules. These have evolved as a result of tacit knowledge and embodied practices, as well as formalised building regulations and ergonomic standards to which manufacturing dimensions and modular prefabrication are codified. BioMateriOME provides an opportunity to observe public interactions with materials and collect detailed microbial analyses from the same materials in parallel. These data might then inform future design conventions with updated knowledge and new material technologies.

As research progresses, it will inform microbially aware design practices, symbiotic spaces and understanding of how microbial communities behave in the built environment. Scientifically, such design has the potential to benefit a given habitat through improved air quality, light levels, acoustic and thermal performance and biocompatibility. Further understanding of microbiomes on home surfaces may lead to the development of novel materials or biotechnologies which could limit the presence of disease-causing (pathogenic) microbes and promote non-pathogenic microbial diversity in indoor spaces, for example, through designing with **probiotic** ↗ tiles (Beckett, 2021).

BioMateriOME opens discussions surrounding cohabiting with living materials, probiotic inhabitation and wellbeing, with opportunities to address these in parallel with design standards so that they can be assimilated into capital as well as operational expenditures in our built environments. Opening dialogues between interdisciplinary teams incorporating microbiologists, designers, architects and building services engineers may serve to facilitate the uptake of microbiological monitoring in the built environment by industry, including sustainable engineers, designers, architects, property developers, and property control and building services departments. BioMateriOME offers an interdisciplinary path towards promoting healthier working and living spaces, which could eventually be formalised and codified through incorporation into either building standards or benchmarks such as BREEAM, LEED or WELL | **Figure 13** ↗ |.

**Figure 13**
Mockup: How might a better understanding of our relationship with the microbiome around us impact on our building standards

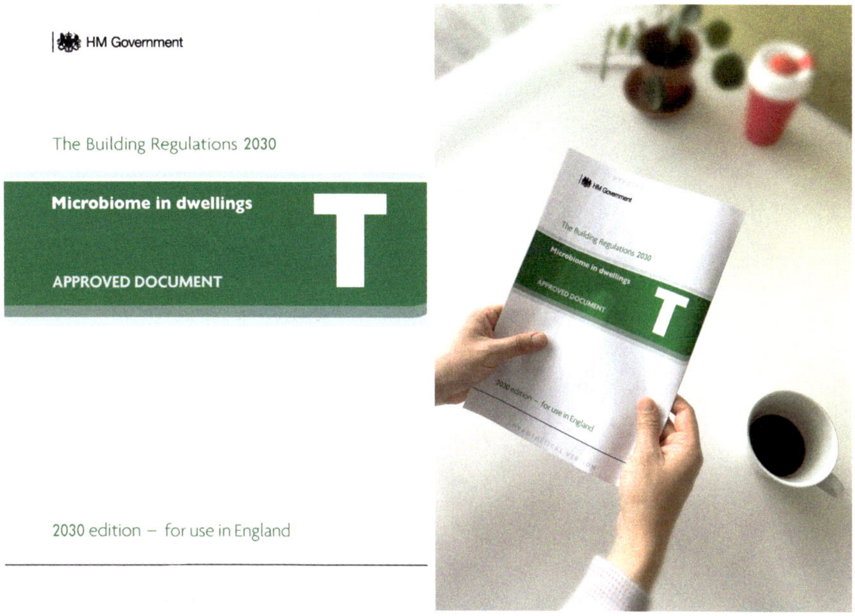


## References


Beckett, R. (2021) 'Probiotic design', *The Journal of Architecture*, 26, 6–31.

Bluyssen, P. M., Janssen, S., van den Brink, L. H., de Kluizenaar, Y. (2011) 'Assessment of wellbeing in an indoor office environment', *Building and Environment*, 46(12), 2632–2640.

Browning, W. D., and Ryan, C. O. (2020) *Nature inside: a biophilic design guide.* RIBA Publishing, London, U.K.

Coughenour, C., Stevens, V., and Stetzenbach, L. D. (2011) 'An evaluation of methicillin-resistant *Staphylococcus aureus* survival on five environmental surfaces', *Microbial Drug Resistance*, 17, 457–461.

Crawford, A., In-na, P., Caldwell, G., Armstrong, R., Bridgens, B. (2022) 'Clay 3D printing as a bio-design research tool: development of photosynthetic living building components', *Architectural Science Review*, 65:3, 185–195.

Dade-Robertson, M. (2021) *Living Construction*. Routledge, Oxfordshire, UK.

Fiorillo, L., Cervino, G., Matarese, M., D'Amico, C., Surace, G., Paduano, V., Fiorillo, M. T., Moschella, A., La Bruna, A., Romano, G. L., Laudicella, R., Baldari, S., Cicciù, M. (2020) 'COVID-19 surface persistence: a recent data summary and its importance for medical and dental settings', *International Journal of Environmental Research and Public Health*, 17, 3132.

Franklin, K., and Till, C. (2018) *Radical Matter: Rethinking Materials for a Sustainable Future*. Thames & Hudson. London, U.K.

Gibbons, S. M., Schwartz, T., Fouquier, J., Mitchell, M., Sangwan, N., Gilbert, J. A., Kelley, S. T. (2015) 'Ecological succession and viability of human-associated microbiota on restroom surfaces', *Applied and Environmental Microbiology*, 81, 765–773.

Gusareva, E. S., Acerbi, E., Lau, K. J. X., Luhung, I., Premkrishnan, B. N. V., Kolundžija, S., Purbojati, R. W., Wong, A., Houghton, J. N. I., Miller, D., et al. (2019) 'Microbial communities in the tropical air ecosystem follow a precise diel cycle', *Proceedings of the National Academy of Sciences of the United States of America*, 116(46), 23299–23308.

Hasling, K. M., Ræbild, U., Herrtua, I., Patel, A. (2020) 'Material Pathways. Narrating materials in design for sustainability (tool)'.

Kampf, G., Todt, D., Pfaender, S., Steinmann, E. (2020) 'Persistence of coronaviruses on inanimate surfaces and their inactivation with biocidal agents', *Journal of Hospital Infection*, 104, 246–251.

Kirjavainen, P. V., Karvonen, A. M., Adams, R.I., et al. (2019) 'Farm-like indoor microbiota in non-farm homes protects children from asthma development', *Nature Medicine*, 25, 1089–1095.

Miodownik, M. A. (2015) 'Towards designing new sensoaesthetic materials: The role of material Libraries', in *The Social Life of Materials*, 1st Edition, 11, Routledge.

Parajuli, A., Grönroos, M., Siter, N., et al. (2018) 'Urbanization reduces transfer of diverse environmental microbiota indoors', *Frontiers in Microbiology*, 9.

Solanki, S. (2020) Connected Living. *Dutch Design Week Talks*. Retrieved from https://ddw.nl/en/video/84/ddw20-talks-connected-living, accessed on 22.10.2020.

Traoré, O., Springthorpe, V. S., and Sattar, S. A. (2002) 'A quantitative study of the survival of two species of Candida on porous and non-porous environmental surfaces and hands', *Journal of Applied Microbiology*, 92, 549–55.

Trubiano, F. (2006) 'Material ConneXion: The Global Resource of New and Innovative Materials for Architects, Artists and Designers and Transmaterial: A Catalogue of Materials that Redefine Our Physical Environment and Material Architecture: Emergent Materials for Innovative Buildings and Ecological Construction – Edited by George M. Beylerian, Andrew Dent, Anita Moryadas, Blaine Brownell and John Fernandez', *Journal of Architectural Education*, 60(2), 72–74.

United Nations (2015) The UN Sustainable Development Goals. United Nations, New York.

Wakefield-Rann, R., Fam, D., and Stewart, S. (2020) 'Microbes, chemicals and the health of homes: Integrating theories to account for more-than-human entanglements', *BioSocieties* 15(2): 182–206.

# 2.6

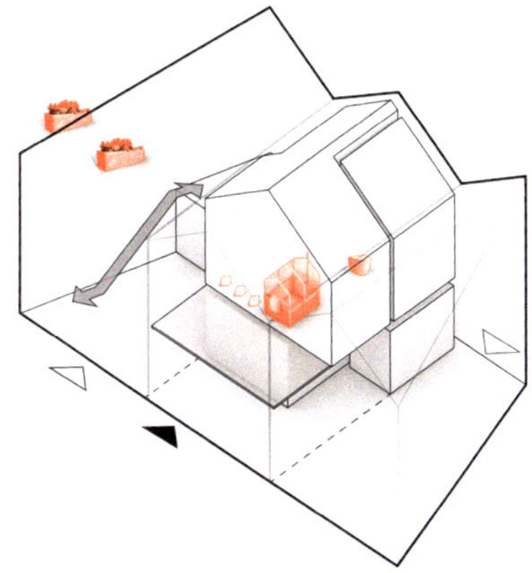

**Figure 1**
Location diagram

# Towards a Self-sustaining Home

## Circular Flows of Materials and Energy in the Domestic Environment

**Data**
Application: Alternative technologies and systems to reduce environmental impact of key building functions including operational energy, food and water consumption.
Building operational energy (heating, cooling, lighting) is responsible for: 17.5 per cent of global $CO_2$ emissions.
Food production and transportation is responsible for: 18.4 per cent of global $CO_2$ emissions.
Proportion of world population without safely managed sanitation: 40 per cent.
Source: https://ourworldindata.org

**Team**
Pippa McLeod-Brown: Architect, Tiny Energy and Oikrobia Concept • Ben Bridgens: Architectural Technologist • Louise Mackenzie: Artist, Participatory Design • Kaajal Modi: Design Researcher • Shafeer Kalathil: Microbial Electrochemist, Microbial Fuel Cell Research • Rajesh Bommareddy: Biotechnologist, Anaerobic Digestion Waste to Energy Research • Gary Black: Protein Biochemist • Oliver Perry: Technical Officer

↗ **Glossary**
ecosystems
microbial fuel cells

There is no waste in nature. Natural **ecosystems** ↗ have evolved over millennia to break down and recycle everything, with no residual waste that cannot return to the system. Biological organisms have evolved to co-exist symbiotically as part of a highly complex network of interwoven systems, feeding off and supplying resources to one another. The only external input in natural systems is energy from the sun. Could the built environment operate in this way, with each house or community being entirely self-sufficient? This chapter takes precedent from nature's zero-waste circularity and approaches the carbon emissions of the home from an operational perspective. It explores how creating circularity between existing biotechnologies and bioprocesses, combined with traditional approaches to making the best use of limited local resources, can reduce both the carbon footprint and waste generated by domestic homes. This can partially be achieved through the production of biogas in a domestic-scale anaerobic digester to reduce natural gas consumption, and the production of electricity through **microbial fuel cells** ↗.

By examining these two technologies we see that a more complex web of processes is required to achieve circularity, and that we must rethink how we use resources in the home to make best use of the small amounts of energy generated by biological systems. The Oikrobia@OME project identifies the power of food production as a generator for circularity in the built environment, linking flows of energy and materials. The potential of Oikrobia@OME is shown through three case studies which demonstrate further uses of material flows within the OME.

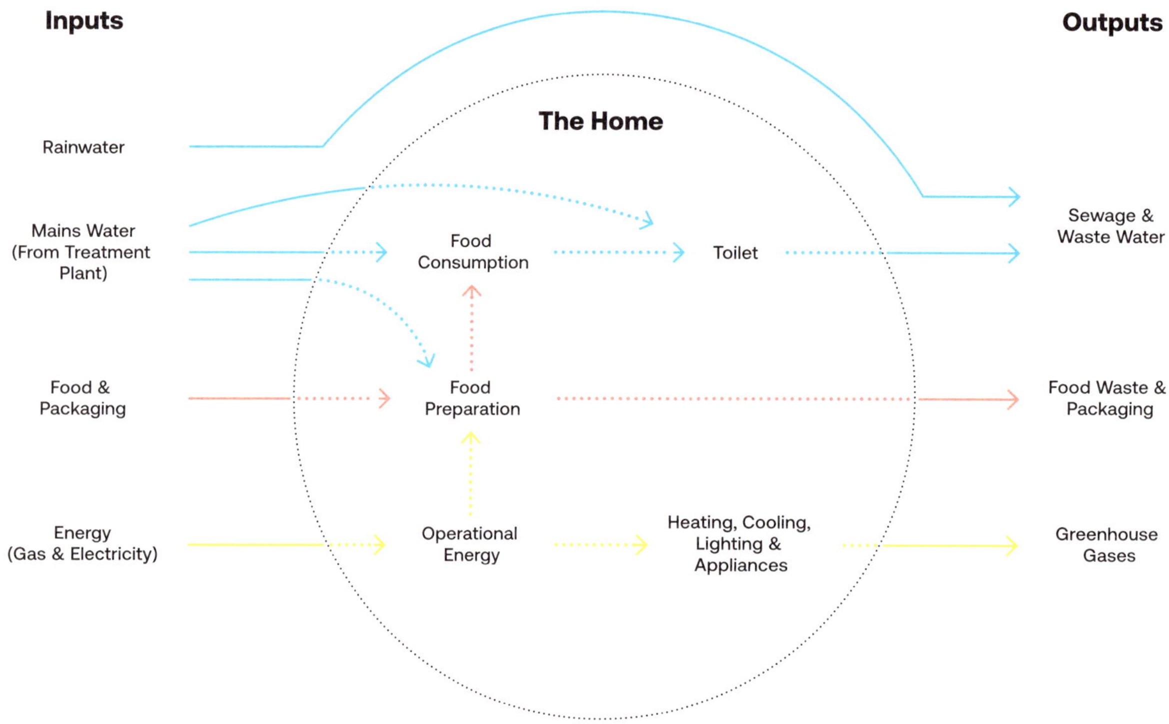


**Figure 2**
Typical domestic linear flows of materials and energy

## Energising Waste: Enabling Technologies for Generating Energy From Domestic Waste Streams

↗ **Glossary**
eutrophication
anaerobic digestion

The current waste treatment system in the UK relies on households separating their waste responsibly; however, the waste collection system is convoluted and complex, with regional variations in policy. Large amounts of organic waste are disposed of in landfill where the lack of oxygen prevents microorganisms from biodegrading it, leading to the generation of methane. In the UK water **eutrophication ↗** is an increasing issue due to the UK's reliance on a combined system where sewage and wastewater are treated together. In periods of intense rainfall the system overloads and untreated effluent is released into rivers. This Victorian sewerage infrastructure has become unsuitable for the increased load from the growing population combined with greater storm intensity due to climate change. In many countries there is no sewage treatment infrastructure, resulting in untreated sewage left in open pits or entering rivers, with severe consequences for human health. The treatment of human waste within the home using biotechnologies to generate energy would benefit both high- and low-income countries. Two key biotechnologies are required to do this: **anaerobic digestion ↗** and microbial fuel cells | **Figure 3 ↗** |.

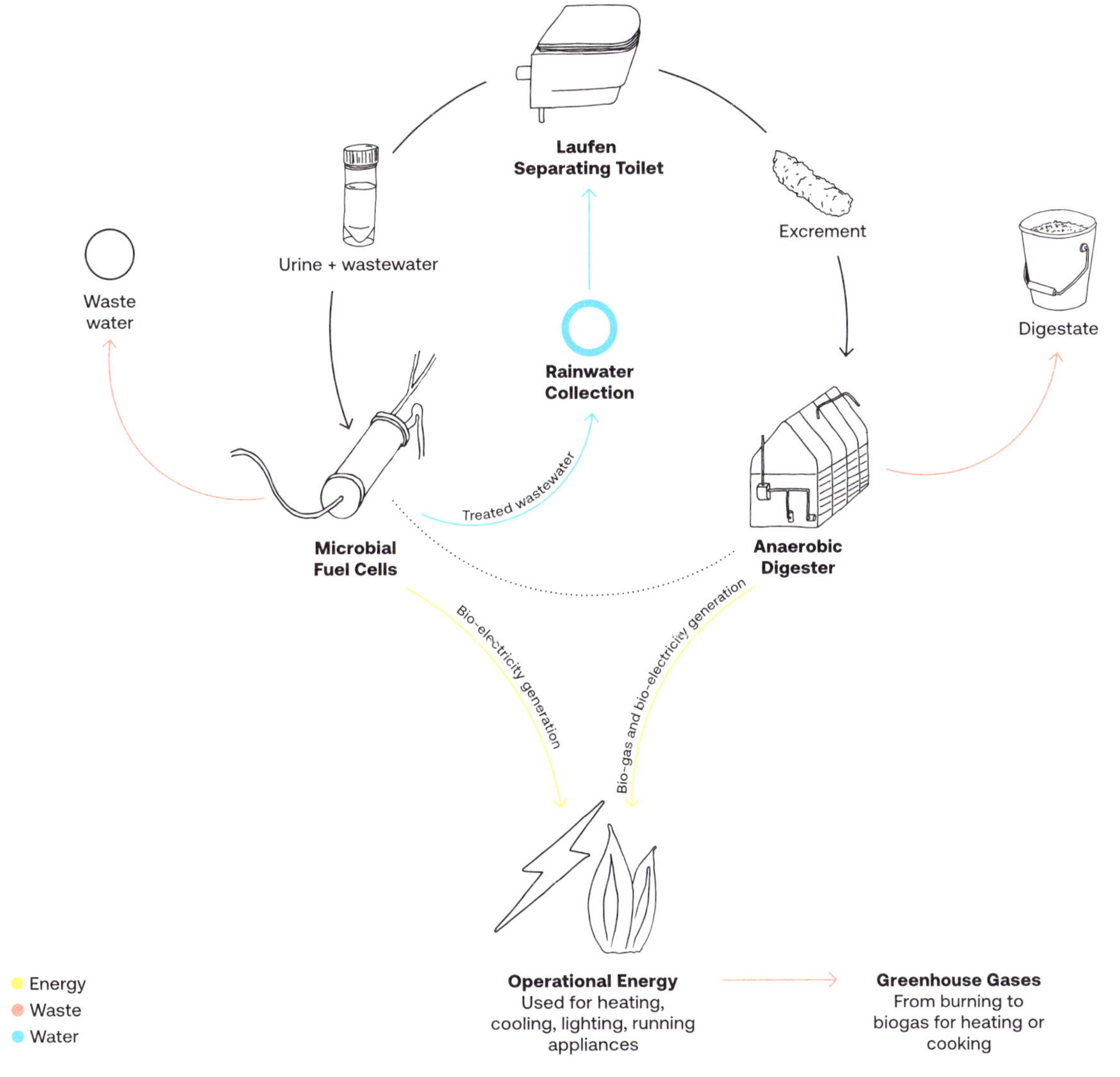


**Figure 3**
Processing domestic waste to generate energy

## Enabling Technology 1: Anaerobic Digestion

↗ **Glossary**
digestate

Anaerobic Digestion (AD) is a natural biological process which occurs in some soils and lake sediments where microorganisms break down organic matter in the absence of oxygen to produce methane. The same process can be used to break down organic waste, such as animal faeces, food waste or agricultural waste, producing biogas and fertiliser | **Figure 4 ↗** |. Biogas is composed mainly of methane, the primary component of natural gas, which is widely used for cooking, heating and large-scale electricity generation. Biogas also contains small amounts of carbon dioxide and other impurities that are removed by separation. The resulting material, called **digestate ↗**, is rich in nutrients and suitable as a crop fertiliser | **Figure 5 ↗** |.

Large-scale commercial AD facilities are widely used to generate energy from food waste, agricultural waste and sewage sludge. This is effective if large quantities of organic matter are produced at a single location (e.g. a farm) or if adequate infrastructure exists to collect organic matter from dispersed sources (e.g. sewage treatment). It is possible to process organic waste in a domestic-scale anaerobic digester. This is popular in rural areas of low-income countries where there is no access to infrastructure for waste processing or energy provision, and in these locations simple, domestic-scale AD units are commercially available (e.g. PUXIN Portable Assembly Biogas System).

## Enabling Technology 2: Microbial Fuel Cells

Microbial Fuel Cells (MFCs) are electrochemical devices that employ bacteria as living biocatalysts to oxidise waste and produce electricity (Teli et al., 2016). A typical MFC is divided into two chambers: aerobic and anaerobic | **Figure 6 ↗** |. In the anaerobic chamber where there is no oxygen, bacteria metabolise waste and electrons generated during this process are transported to the electrode (anode), while oxygen in the aerobic chamber receives electrons by combining with hydrogen to form water. This flow of electrons generates electricity.

Power output from MFCs is still too low to run large devices or equipment. However, MFCs have been successfully employed to charge small devices such as mobile batteries by stacking multiple MFC units together. Even though an MFC produces low current, it is a compelling technology to treat wastewater compared to conventional approaches. MFCs generate less sludge, are tolerant to salinity and changes in pH, and can be operated at room temperature.

## Enabling Technology 3: Separating Toilet

Large-scale, centralised wastewater treatment systems process a mixture of human faeces and urine, domestic wastewater and rainwater. This greatly simplifies collection and the water is required to help transport the faeces. However, it is not the most efficient or effective way to process each waste component. Faeces are well suited to anaerobic digestion; urine can be processed in a microbial fuel cell (with as little water as possible to give the highest energy yield); and it is more efficient to collect water and reuse it locally. A separating toilet avoids mixing or diluting faeces and urine, and can deliver them to AD and MFC systems respectively (Otterpohl, Braun and Oldenburg, 2004). Separating toilets are usually 'dry' and are regarded as being smelly and requiring more maintenance than a conventional flushing toilet, hence they are not common across most domestic settings.

Toilets are culturally sensitive and any change to accepted toilet practices are often cited as a barrier to implementation (Ahmed and Ahmed, 2017). A recent development by the Swiss company Laufen aims to eliminate these cultural barriers to localised processing of human waste. The Laufen Save! toilet uses water to contain the smell of the faeces in the same way as a normal toilet. Surface tension is used to divert the urine into a separate waste pipe, while the faeces fall into a conventional wet flush toilet which is served by a second waste pipe | **Figure 7 ↗** |. The urine is undiluted except for a small volume of water in a U-bend to avoid odour and meet regulations. The faeces are mixed with water but for AD dilution is less significant.

Using a wet flush system enables the faeces to be transported along a pipe to the anaerobic digester, unlike a dry system in which the processing chamber must

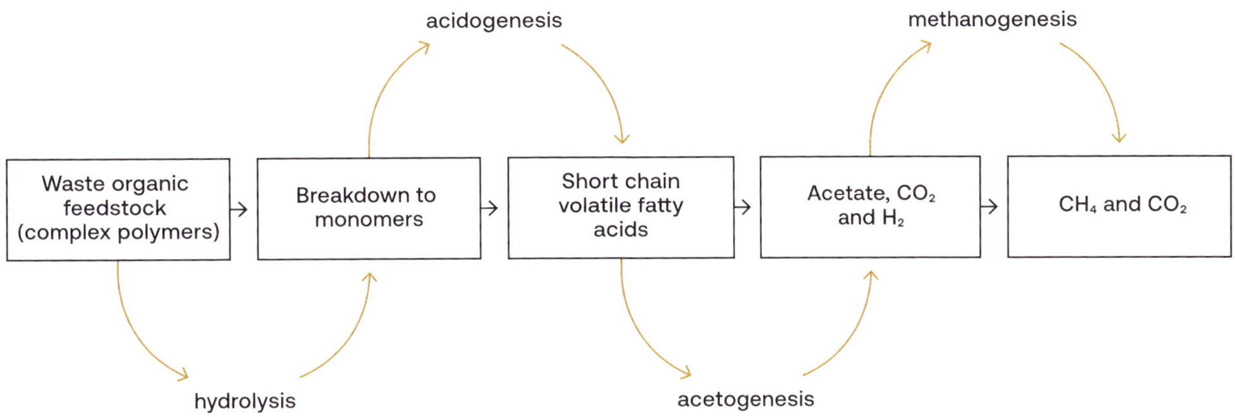


Figure 4

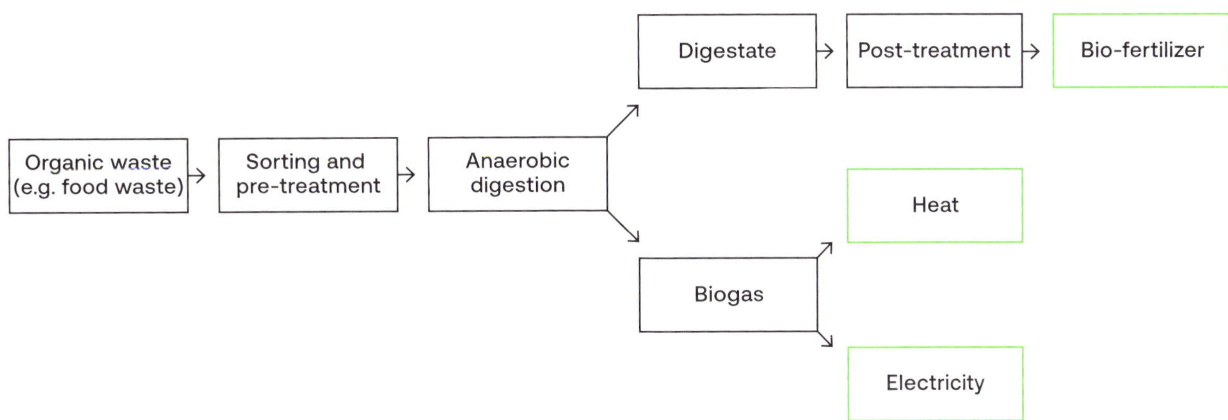


Figure 5

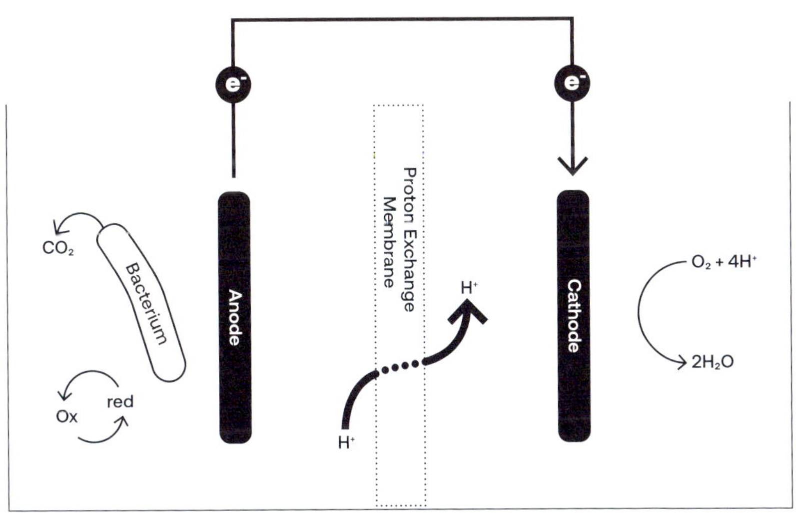


Figure 6

**Figure 4**
The four phases of bacterial degradation of organic waste producing biogas in an anaerobic digestion process

**Figure 5**
Example of an AD configuration producing energy and bio-fertiliser from organic waste

**Figure 6**
Microbial fuel cell principle of operation

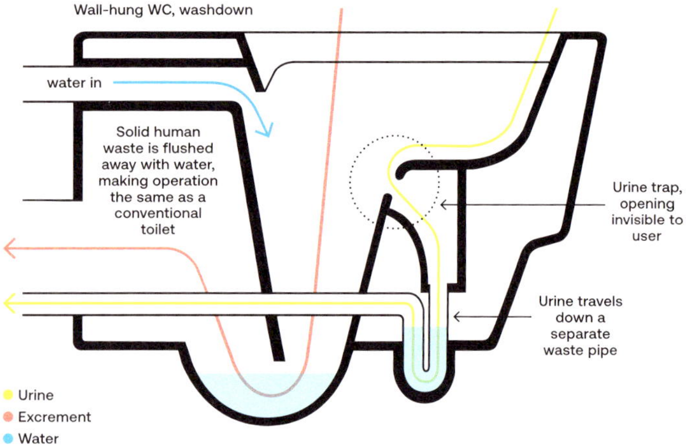


Figure 7

Figure 8

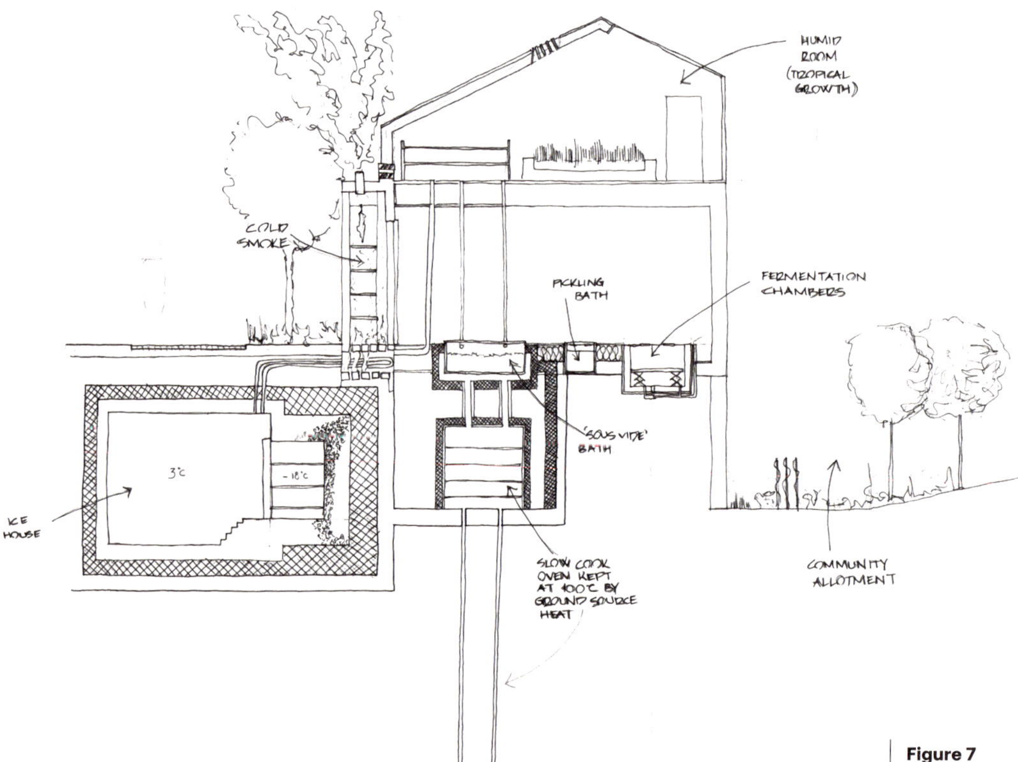


Figure 9

**Figure 7**
Laufen separating toilet diagram

**Figure 8**
Laufen separating toilet

**Figure 9**
Architectural design driven by domestic low energy food preparation

be situated directly below the toilet. This is very significant for integration in domestic buildings, especially for retrofit of existing buildings where the anaerobic digester could be situated outside or in a garage or utility room. The result is a toilet which is indistinguishable from other modern European toilets and can be fitted in the same space | **Figure 8 ↗** |. The only extra requirement in use is that for urine diversion to work, the user must sit on the toilet, not urinate while standing.

## Oikrobia: Developing Circular Flows in the Home

↗ **Glossary**
ecosystems

The enabling technologies described above have the potential to generate energy locally from human and food waste at a domestic scale whilst eliminating the need for sewage transport infrastructure and treatment facilities. However, these technologies have two limitations: they generate relatively small amounts of energy, and they do not create a closed-loop, self-sustaining system by themselves. Natural **ecosystems ↗** are highly complex with many organisms that endlessly recycle energy and nutrients. Similarly, a more complex system can help achieve circularity in the home, harnessing energy from the sun and ensuring that flows of nutrients can be cycled.

Food can play a central role in this more complex system. In the last century food has transitioned from a locally produced, carefully managed resource to a globally traded commodity of which 30 per cent is wasted and is responsible for 25 per cent of global greenhouse gas emissions (IPCC, 2014). As the digestate from an anaerobic digester is an excellent fertiliser, it seemed natural to expand the scope of the system to include food production.

## Tiny Energy

After conducting preliminary research into bacteria spore-based hygromorphs (further discussed in Chapter 3.3), Pippa McLeod-Brown devised the Tiny Energy concept. Whilst considering the micro-scale power outputs of these materials, McLeod-Brown speculated on how they could become tangible amounts of power. As Jane Bennett discusses in *Vibrant Matter* (2010), there is an imperceptibly small amount of energy inherent in all matter that is cumulatively a force so strong it governs the order of the universe. Tiny Energy builds on Bennett's theories in an architectural context by defining a biologically stimulated design paradigm which utilises the untapped potential of vibrant matter for ecological potentialities (McLeod-Brown, 2020a). Humans, however, have become accustomed to changing the properties of natural resources to suit their needs by removing their lively complexity in order to make them more ordered, predictable, and controllable. Tiny Energy encourages designers to efficiently harness the power sources found within nature by working with living systems rather than against them.

## Conceptual Prototype: From Tiny Energy to Oikrobia@OME

↗ **Glossary**
economy

The Oikrobia concept was developed to challenge anthropocentric behaviours and encourage a shift to design thinking that privileges non-human agency. This began as an architecture design thesis, which established a methodology in which the Tiny Energy principles could be applied. Oikrobia is derived from the Greek *oikos + mikros + bios*, the home + small + life. Using food as a generator for circularity in the domestic environment, Oikrobia explores how food can influence the **economy ↗** of the home. Economy comes from the ancient Greek word *oikonomia* meaning household management. We refer to economy in the traditional sense which encapsulates the management of the home in its entirety, as opposed to purely financially. A well-run household would waste nothing and therefore exhibit good economics.

In *Hungry City* (Steel, 2013) and *Sitopia* (Steel, 2020), Carolyn Steel proposes that in order to create more sustainable food systems we must design cities in ways that facilitate circular food production. These compelling arguments influenced the food-based focus of Oikrobia. We know that human civilisations co-evolved alongside the food systems that feed them; the expansion of settlements into today's vast cities was only possible once agricultural and industrial revolutions enabled the production and transportation of larger volumes of food. Yet we now realise that the industrial production and transportation of food is one of the primary drivers of climate change and biodiversity loss. Oikrobia is predicated on the principle that in order for the built environment to become more sustainable, so too must the food production, distribution, consumption and waste systems that underpin its design. The entanglements between food and buildings make food production and consumption an ideal context in which to test the Tiny Energy concept.

In 'Oikrobia: a Microbially Driven Architectural Investigation into a Reformed Food System', food is explored through a microbial lens. The relationship between food and space was interrogated using microorganisms as contributors to architectural futures | Figure 9 ↗ |. In a move away from conventional cooking methods, the research explored microbial fermentation processes that change the texture, flavour, and longevity of foods. These processes led to the development of low-energy cooking techniques that incorporate spatial and environmental parameters for microorganisms, thus informing the human-scale architectural intervention in which Tiny Energy could be implemented (McLeod-Brown, 2020b). Oikrobia has since been used as the foundation of the conceptual Culina GastroLab (Dy, Riley and McLeod-Brown, 2020) for the BioDesign Challenge 2020 and Oikrobia@OME.

## Oikrobia@OME: Closing the Loop

Oikrobia@OME tests the principles of Oikrobia at a domestic scale in the HBBE's experimental house, the OME, by mapping flows of materials and energy, identifying gaps in their circularity, and proposing sustainable ways of closing these gaps to create a holistic system | Figure 10 ↗ |. The OME has a separating toilet installed in the first-floor apartment, with two waste pipes carrying urine and faeces into a small laboratory, in which there is a one-cubic-metre anaerobic digester and bench space for microbial fuel cells. There are exposed services which allow, in due course, gas from the AD to be piped into the kitchen for cooking, and low voltage wiring to be installed, to make efficient use of the electrical supply from the MFCs. There is also external space for raised beds to grow food.

Anaerobic digestion and MFC technologies produce digestate as a 'waste' product and generate relatively small amounts of energy compared to the profligate use of energy to which people are accustomed. Oikrobia proposes the introduction of domestic food production and low-energy cooking to make efficient use of these resources and create circularity. An urban garden can use the digestate as an organic fertiliser due to its nutrient-rich composition. Growth of food in an urban environment not only provides sustenance, but dramatically reduces the negative environmental impacts from industrialised agriculture, transportation and packaging. To make the best use of Tiny Energy from biological processes, traditional low-energy food preparation techniques are suggested – including slow-cooking at low temperatures, smoking, pickling and fermentation (a biological process which requires no external energy input). Waste organic matter from food production and preparation, combined with human waste, is returned to the AD and MFC, and the cycle continues.

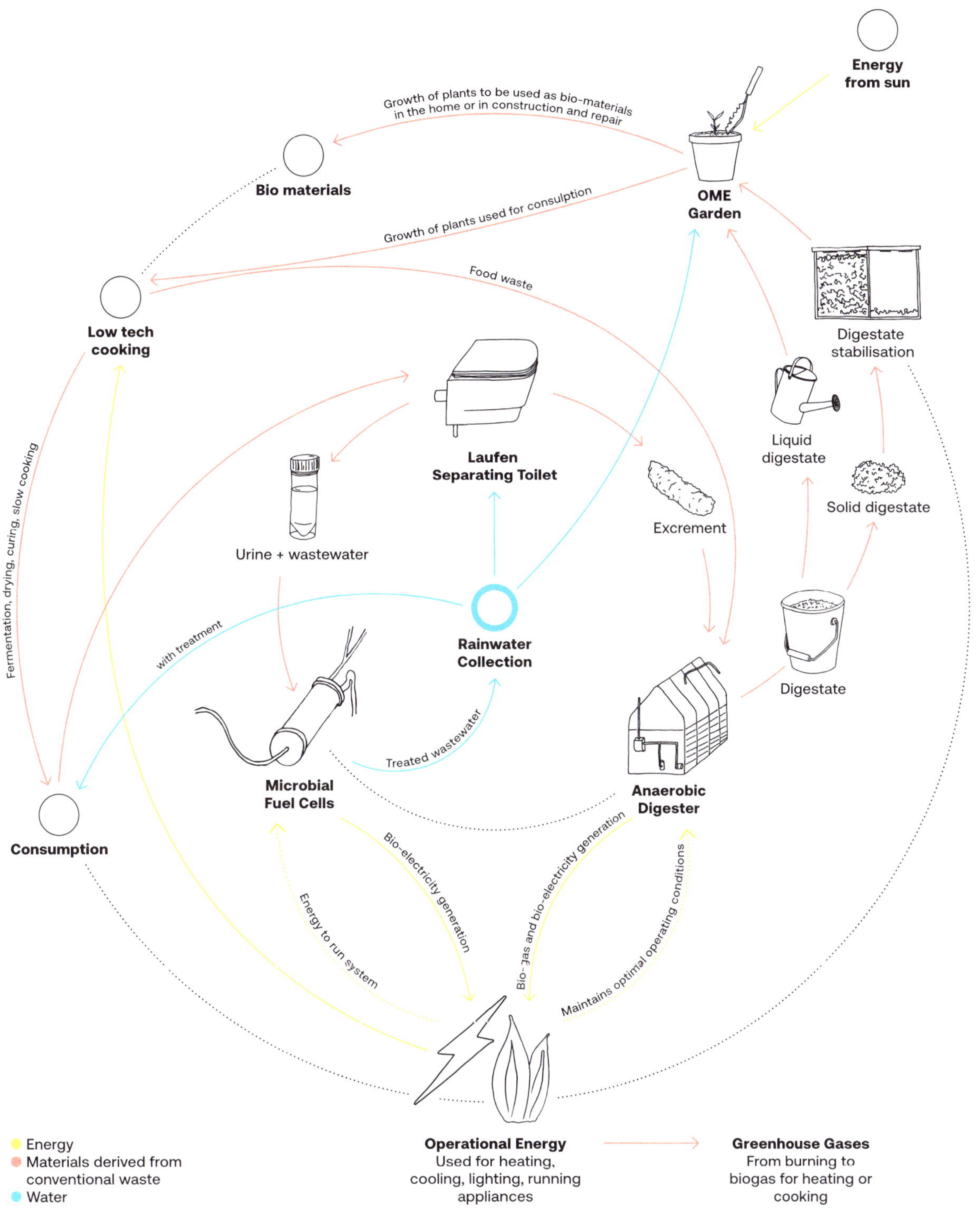


**Figure 10**
Proposed domestic circular flow of materials and energy

## Oikrobia@OME: New Directions

In summer 2022, as part of the Newcastle University Climate Leadership Scholars programme, three Masters of Architecture students were asked to identify a waste stream within the Oikrobia@OME system, and to spend eight weeks investigating whether a biotechnology or biomaterial could provide a way to utilise that waste stream and create greater circularity.

### Project 1

**Von Fabric: A Biomaterial Made From Leek Waste, Iulianiya Grigoryeya**

Iulianiya's interest in woven fabrics led her to investigate fibrous food waste with the ability to be woven into threads. Alliums, the vegetable family that includes onions, garlic and leeks, are a consistent contributor to domestic food waste due to the practice of removing layers of skin before preparing for consumption. As leeks account for 10 per cent of outdoor vegetables grown in the UK, their popularity and the larger surface area of their skin peelings makes the leek a suitable candidate for further investigation. The project developed and trialled a method for drying and rehydrating leek peelings to increase the material's flexibility before testing three different weaving methods | Figure 11 ↗ |. The performance of the woven samples was tested against natural fibres used in construction such as hemp, wool and silk. A public workshop conducted in collaboration with the Great North Museum gathered knowledge on historical and contemporary weaving techniques, and elicited feedback on public perception of the leek-based woven material. This has led to the potential for development of the woven fabric into larger pieces and with other forms of food waste.

### Project 2

↗ **Glossary**
Kombucha Method

**Food Waste Recipes for Bacterial Cellulose Growth, Roxana Caplan**

Bacterial Cellulose (BC) is a biomaterial which is straightforward to grow and has the potential for application in the built environment (Chapter 2.2). Roxana's research investigated the potential for food waste to be used as the nutrient source in (BC) production, and documented the resilience of BC when produced in domestic, and therefore unsterilised, environments. Using the **Kombucha Method** ↗, the nutrient sources tested were common household food wastes with a relatively high sugar content: banana peel, orange peel, apple and pear cores, pineapple skin and watermelon rind | Figure 12 ↗ |. Preliminary tests successfully showed BC growth from media with the identified food wastes providing the carbon source. Following this, experimentation was undertaken to test variation in BC growth between recipes, altering fruit waste quantities to provide approximately 5 per cent and 8 per cent sugar content, and the addition of tea waste as a nitrogen source. Use of low-energy preparation methods in an uncontrolled domestic environment demonstrated the resilience of BC growth and accessibility of the method to the public. The work provides recipes and methods which can be developed to test other food waste types with minimal risk of contamination for consistent BC growth and is a step towards the local conversion of food waste into building materials.

Figure 11

Figure 12

**Figure 11**
Von Fabric images. Process image from the public workshop (top right), leek weaving samples (top left and bottom left)

**Figure 12**
Bacteria cellulose grown using waste food as a nutrient source

## Testing Waste Kombucha as Fertiliser, Ollie Spur

This research project sought to explore the potential use of the liquid produced from Bacterial Cellulose (BC) production using the Kombucha Method as a fertiliser. As mentioned previously, BC is being investigated as an architectural façade material (Chapter 2.2). When BC is grown at architectural scale, kombucha is produced in abundance. Kombucha is a nutrient-rich medium containing nitrogen and carbon, so this project tested the potential of kombucha as a fertiliser by documenting the growth of wheatgrass with and without it. Kombucha was added to the soil with the intention that the nutrients in the solution would enrich the bacterial colonies in the soil and improve the growth of the wheatgrass. Wheatgrass was chosen due to its fast growth time, making this project possible in the short eight-week duration of the scholarship | Figure 13↗ |. The project needs more time to ascertain the value of kombucha as a fertiliser, but an unexpected outcome of this research was that BC formed around the roots of the wheatgrass, creating a BC sheet reinforced with roots. This suggests further ways that growth and waste materials could be combined to create useful products.

**Figure 13**
Using kombucha as fertiliser. Kombucha (top left), wheatgrass growing behind the south facing polycarbonate façade in the OME (top right), wheatgrass germination (bottom left)

## Conclusion – Architectural Integration

↗ **Glossary**
feedstock

Even within the OME, which was designed for testing new technologies, there are challenges for implementation of the AD and MFC technologies and the wider Oikrobia@OME system. Gas safety regulations make it difficult to use an unconventional supply of gas: the current heating system operates with an electric heat pump meaning that biogas cannot easily be used to heat the building; indoor and outdoor space for food production is minimal; an intermittent supply of human waste makes it difficult to run the AD efficiently; the small kitchen offers limited space for unconventional food preparation; and men must be persuaded to sit whilst urinating.

Within existing housing, the challenges are much greater: where can the AD and MFC be situated? A small domestic AD unit requires a one-cubic-metre gas storage bag | Figure 14 ↗ |, and it must be able to receive human waste from the toilet. How much interaction and maintenance do these processes require? How can the biogas and low voltage electricity be effectively used? Is there external space, expertise and willingness to grow food? We conclude by describing recent and future developments which will start to address these challenges.

Commercial AD is carried out at between 25 and 37 °C to maximise the rate of breakdown of organic matter (McKeown et al., 2012). Whilst the process can usefully operate at slightly lower temperatures (e.g. 20 °C), below this temperature the process slows dramatically meaning that AD units must either be heated or located indoors, taking up valuable indoor space in colder climates. Current research is focused on finding natural or engineered microorganisms that can carry out AD at much lower temperatures, which would be significant for enabling the use of external AD units in colder climates. This would make retrofit easier, with installation on rooftops or in gardens, similar to the installation of air-conditioning units or heat pumps. Other research goals include increasing methane yield and ensuring that the AD process can cope with changes in **feedstock** ↗ (Meegoda et al., 2018), necessary for a domestic system which will process a range of waste sources.

Domestic food production is less likely to gain widespread acceptance if it requires substantial external space and considerable effort and skill. There is great potential for recent advances in large-scale vertical urban farming using low-energy LED lights and hydroponic nutrient delivery to be applied at domestic scale, so that walls in the house would be covered with fruit and vegetables growing in ideal conditions with minimal intervention. Alternatively, in less densely populated areas, or for people who wish to engage in food production and eat seasonal produce, food could be grown externally. The larger question remains what is required to support the cultural adoption of such practices, and whether they can be scaled to produce a significant proportion of the food required.

The use of microbial processes to generate fuel and electricity from domestic waste at household scale has the potential to provide a reliable, sustainable, localised energy source without the need for new network infrastructure, whilst reducing the need for domestic waste collection and disposal. Expanding this system to incorporate food production and preparation starts to close the loop with circular flows of energy and materials, driven by energy from the sun. However as the system becomes more complex it begins to intrude on every aspect of domestic life, questioning how we cook and eat, what we must do to maintain such systems and how our houses might be redesigned to accommodate new metabolic processes and habits. A major cultural shift will be required to embrace this new way of living, but the benefits would be enormous – radically reducing environmental impacts in almost every industrial sector.

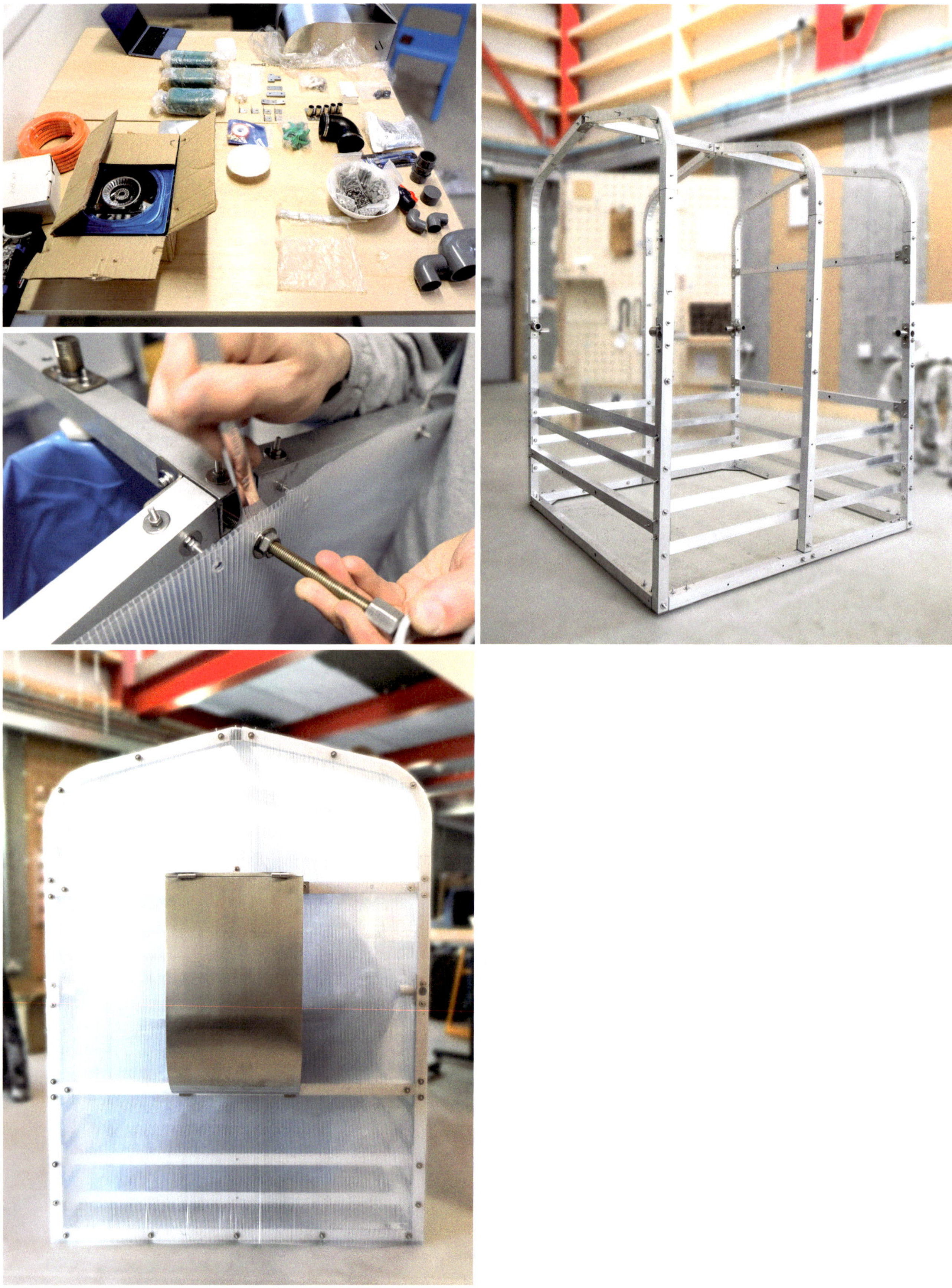

**Figure 14**
Assembly of a one-cubic-metre domestic anaerobic digester, most of the space is used for a gas storage bag which is contained within the metal frame and protected by polycarbonate sheeting

## References


Ahmed, S. K., and Ahmed, S. S., (2017) 'Socio-cultural acceptability of urine diverted composting toilets: A review of literature for possible adoption in peri-urban areas as a sustainable sanitation solution', in *AIP Conference Proceedings* (Vol. 1919, No. 1). AIP Publishing LLC.

Bennett, J., 2010. *Vibrant Matter: A Political Ecology of Things*. Duke University Press.

Dy, A., Jeong, D., Riley, E., McLeod-Brown, P. (2020) *Culina Gastro-Lab: The Future of Food (Biodesign Challenge 2020)*. https://www.youtube.com/watch?v=uO1t4zyHAxA&t=349s

IPCC (2014): Climate Change 2014: Synthesis Report. Contribution of Working Groups I, II and III to the Fifth Assessment Report of the Intergovernmental Panel on Climate Change [Core Writing Team, R. K. Pachauri and L. A. Meyer (eds.)]. IPCC, Geneva, Switzerland, 151.

McKeown, R. M., Hughes, D., Collins, G., Mahony, T., and O'Flaherty, V. (2012) 'Low-temperature anaerobic digestion for wastewater treatment', *Current Opinion in Biotechnology*, 23(3), 444–451.

McLeod-Brown, P. (2020a). 'Hygrospores: Material Investigation into a Tiny Energy for Architecture', Newcastle University

McLeod-Brown P. (2020b) 'Oikrobia: A Microbially Driven Investigation into a Reformed Food System', Newcastle University.

Meegoda, J. N., Li, B., Patel, K., Wang, L.B. (2018) 'A review of the processes, parameters, and optimization of anaerobic digestion', *International Journal of Environmental Research and Public Health*, 15(10), 2224.

Otterpohl, R., Braun, U., and Oldenburg, M. (2004) 'Innovative technologies for decentralised water-, wastewater and biowaste management in urban and peri-urban areas', *Water Science and Technology*, 48(11–12), 23–32.

Steel, C. (2013) *Hungry City: How Food Shapes Our Lives*. Random house.

Steel, C. (2020) *Sitopia: How Food Can Save the World*. Random House.

Teli, N. C., Bhalerao, S. A., Didwana, V. S., Verma, D. R. (2016) 'Microbial fuel cell: a source of sustainable energy', *BIOVISTAS International Journal of Biological Research*, 5(6), 1–12.

Ruth Morrow
Ben Bridgens
Louise Mackenzie

3

# MicroAssemblies
## Benchtop Biotechnological Prototypes

↗ **Glossary**
Prototypes

In this section, we focus on inspirational concepts and practices that indicate the future of biotechnology in the built environment. A series of small-scale 'benchtop' **prototypes**↗ illustrate the potential that working with biological material has to offer the spaces in which we dwell. These microassemblies are all concepts which have only been tested at the most experimental of levels. Some are still lab-based, and others have in small ways been brought into community contexts. All show tantalising possibilities of a future in which biology appears to offer us options for more sustainable, multispecies ways of living.

Designers, architects and scientists who have first-hand experience of working with microbial materials share how they have learned to work with organisms that cannot be controlled, but might instead be cajoled into acting in a way that is useful to us, whilst also being mindful of the needs of the organisms themselves. These authors paint a picture for us of the challenges that lie ahead in the literal 'domestication' of biological resources, whilst also inspiring us to consider new ways in which we can learn from living organisms and work alongside them.

We learn how microscopic algae that are found naturally in water-rich environments might exist in a dormant state on dry materials and be revived, when needed, to act as urban carbon sinks. We learn how a combination of biological materials and biological mimicry can create smart materials that change their form according to external parameters such as heat, moisture and light. We consider the possibilities that exist when we allow biological organisms to express their own forms within a given set of parameters, whether those be environmental, in the case of fungi growth, or capacity-based, in the case of microbial forms of cement. And we consider the possibilities for local-scale production of products including medicines, fuel and food that could exist as DIY-Bio communities come together to make use of difficult-to-recycle waste streams.

In this section it is interesting to note that all of the authors identify the complexities inherent in moving to the next stage of development. Not only must we find ways to work with (rather than employ) biological organisms, but we must also find ways to work with a greater range of human partners than perhaps ever before to fully comprehend how these seeds of ideas can grow into physical, meaningful futures. As one author suggests, the DIY community may have the breadth of skills and the dexterity to take these ideas forward. The challenge that remains is whether the wider societal 'organism' will be flexible enough to adapt.

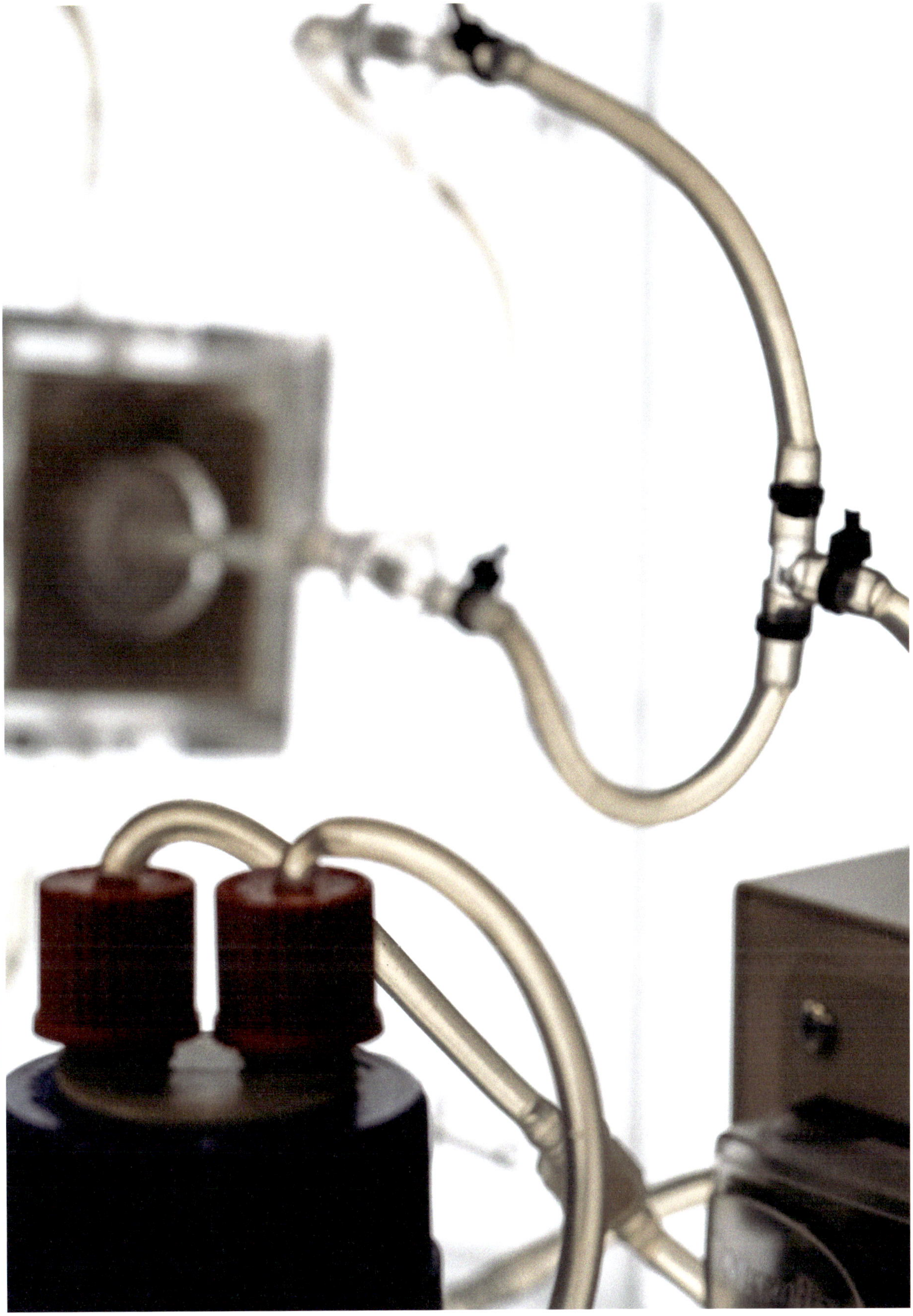

**Figure 1**
A benchtop bioreactor provides
the conditions necessary for life

# 31

## Thora Arnardottir

# 3.1

# Bacterial Sculpting

## Customising Biofabrication Techniques for Biomineralisation

### Context

↗ **Glossary**
calcareous
biomineralisation
Microbially Induced Calcium Carbonate Precipitation (MICCP)
biocement

Microorganisms are ever present in soils and sediments, where they play a vital role in a range of biological processes. They can produce hard **calcareous** ↗ materials with little energy expenditure. One such example is microbial **biomineralisation** ↗: a process that occurs in common soil bacteria, *Sporosarcina pasteurii*, and which is capable of binding sand-sized particles into materials like sandstone with minimal energy input and at room temperature | Figure 1 ↗ |. The process is also known as **Microbially Induced Calcium Carbonate Precipitation (MICCP)** ↗ and has been widely explored in ground engineering to improve soil stability (Van Paassen et al., 2010); the production of **biocement** ↗ (Whiffin et al., 2007) and bricks (Dosier, 2011); and in the repair of cracks in self-healing concrete (Jonkers et al., 2010).

Through MICCP, we can control certain environmental factors by generating optimal conditions for the bacteria to grow and for biomineralisation to occur. Even though the context of this work is situated in the built environment, the primary interest was not in creating new construction materials but in the co-designed emergence of form. I therefore designed and built bespoke bioreactor-casting vessels | Figure 1 ↗ | that allowed me to engage in a process of material tinkering, where I enabled matter to form through biological processes. By combining microbiology and design, the creation of these vessels accompanied a study of how we can alter our fabrication approaches to include the living system.

**Figure 1**
Microscopic image of a Bacterial colony (*Sporosarcina pasteurii*) grown on urea and calcium chloride enriched nutrient agar with precipitated calcite mineral spheres

Figure 2

Figure 3

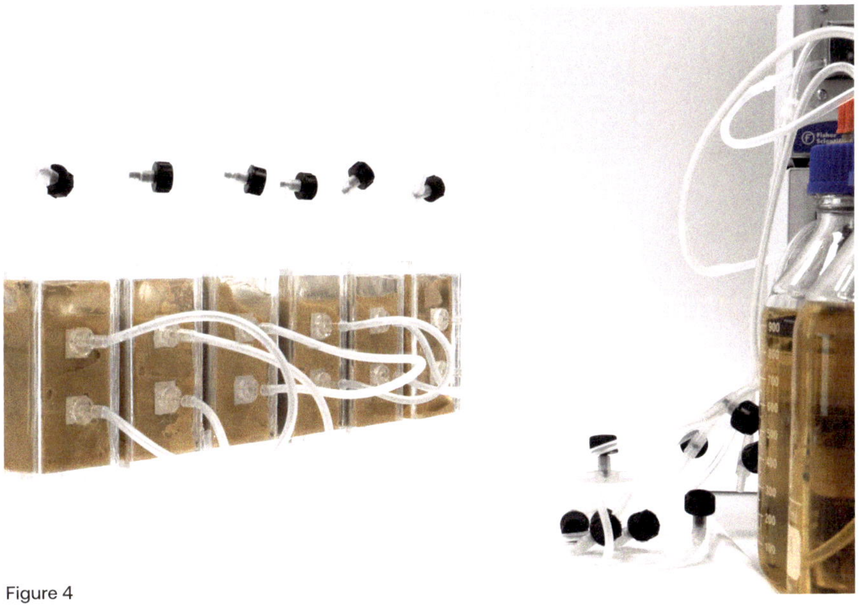

Figure 4

**Figure 2**
One of the bioreactor-casting vessels for a cube mould suspended inside a bioreactor. The bioreactor allows fluid media to be pumped in from multiple inlet points during the casting process

**Figure 3**
Assembled acrylic cube mould filled with composite sand and bacteria mixture

**Figure 4**
Laboratory set-up, exploring flow as a catalyst for the biomineral formation

## Issues and Lessons

↗ **Glossary**
prototype
nutrient media

Many interconnected influences affect the MICCP process. The Bacterial Sculpting **prototype ↗** shown here explores the shape and size of the casting vessel, the size of the granular sand material, and the consistency and frequency of the fluid **nutrient media ↗**. Throughout this exploration, the influence of the flows of nutrient media became apparent in the system. Initial experiments proved that the inlet locations where nutrient media were pumped into the vessel provided the strongest biomineralised bond. However, this also blocked the media from accessing the rest of the volume, causing inconsistencies in the sample. To respond to those challenges, inlet and outlet points for the flow of the media were placed on all sides of the casting volume | **Figure 3 ↗** |.

Further flow exploration was conducted by injecting the media from the inside of the volumes | **Figure 4 ↗** | and allowing the media to flow out towards the sides during casting. The results from these tests challenged the idea that, in casting, the material takes the form of the cast (Arnardottir et al., 2021) and rather offered an exploration into form-finding that revealed the heterogeneity of flows hidden in the casting process.

## Conclusions and Future Potential

↗ **Glossary**
biocomposite

Rather than imposing a precise shape, the bespoke vessels allow the exploration and capture of this complex living process with potential to form hard calcareous materials. The design of these vessels is aimed at exploring material and fabrication methods that incorporate living cells as an inherent part of the fabrication process whilst providing parameters that facilitate the synthesis of the **biocomposite ↗** material resulting from their activity. The vessels simultaneously supply the bacteria with nutrients to sustain their growth and induce a reaction that changes their chemical environment. This precipitates the production of calcium carbonate crystals binding the sand together.

I define this process as *unruly* because its behaviour is not based on the rules we would expect it to follow or impose upon it. Instead, the matter follows and reacts to the variables in its environment and shows subtle signs of its tendencies in relation to them. The research looks at unruly fabrication as a practice that lingers between the designed and the biological. The designer steers the process only up to a certain point: ultimately, the organism's response controls the emergence of the final form.

While containing consistent features, each cast is unique and represents a story of the living, dynamic processes it contains | **Figure 5 ↗** |. This situates the designer as a facilitator of the process: not in complete control of the material properties, but as a coordinator of influential factors that bring about the biomineralised material.

This work is a part of the PhD thesis titled 'Bacterial Sculpting: a processual approach to forming with unruly matter'.

**Figure 5**
Result samples of biomineralised cubes

## References


Arnardottir, T., Dade-Robertson, M., Mitrani, H., Zhang, M., Christgen, B. (2021) 'Turbulent Casting: Bacterial Expression in Mineralized Structures', in Slocum, B., Ago, V., Marcus, A., Doyle, S., Yablonina, M., and del Campo, M. (eds.) *ACADIA 2020 Distributed Proximities: Proceedings of the 40th ACADIA Annual Conference*, Vol. I., 300–309.

Dosier, G. K. (2014) *Methods for Making Construction Material Using Enzyme Producing Bacteria*. (United States Patent No. US 20140239535 A1). United States Patent and Trademark Office. https://ppubs.uspto.gov/pubwebapp/

Jonkers, H. M., Thijssen, A., Muyzer, G., Copuroglu, O., Schlangen, E. (2010) 'Application of bacteria as self-healing agent for the development of sustainable concrete', *Ecological Engineering*, 36 (2), 230–235.

Van Paassen, L. A., Ghose, R., van der Linden, T. J. M., van der Star, W. R. L., van Loosdrecht, M. C. M. (2010) 'Quantifying biomediated ground improvement by ureolysis: Large-scale biogrout experiment', *Journal of Geotechnical and Geoenvironmental Engineering*, 136 (12), 1721–1728.

Whiffin, V. S., van Paassen, L. A., and Harkes, M. P. (2007) 'Microbial carbonate precipitation as a soil improvement technique', *Geomicrobiology Journal*, 24 (5), 417–423.

# 3.2

Emily Birch

# 3.2

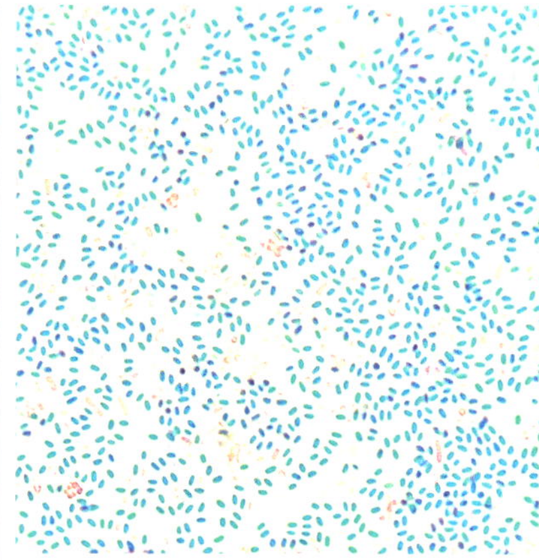

**Figure 2**
Microscopic image of Bacillus subtilis, stained to highlight viable spores (blue), cell debris and live cells (red)

# Bacterial Hygromorphs
## Harnessing Moisture-sensitive Biodynamics Into Responsive Smart Materials

## Context

↗ **Glossary**
morphology
hygromorphic
prototype
adhesion
deposition
biomaterial

There is an incredible tenacity and resilience within the natural world to adapt to changes in the immediate environment. Evolution has refined form to follow function so elegantly that perceivably inert structures such as the pinecone can respond effortlessly to environmental changes. Their unique tissue **morphology**↗ engineers the motion required for seed dispersal even after death. If we could utilise such systems within architectural design, our reliance upon mechanical and fossil-fuel systems to maintain our internal environments could finally be made redundant. The urgency for such solutions has been growing in response to the climate crisis, and the depletion of finite resources has propelled the development of smart materials to the forefront of scientific research (Koyaz, 2018; Lendlein et al., 2019). Can we develop smart materials which replicate nature's adaptability for application within the built environment?

**Figure 1**
Investigations into folding and self-assembly

**Hygromorphic ↗** materials respond to environmental humidity by swelling in humid conditions and shrinking in dry environments, passively, without energy input. A common soil bacterium (*Bacillus subtilis*) develops robust, desiccated spores in hostile environments. By utilising the spores' natural hygromorphic properties within a bilayer (a layer of hygromorphic spores adhered to an inert layer of latex sheet) we can harness their properties of expansion and contraction into a directional response. We describe this as a bioactive hygromorphic actuator ('actuator'). This potential was explored in the study 'Active Origami', where a fabrication method was developed to produce these actuators spore culture methodologies were optimised | Figure 2 ↗ | and simple programmability of the actuators was investigated through altering the concentration of spores used to form the bilayer | Figure 3 ↗ | (Birch et al., 2021).

Using a multidisciplinary approach, a synergy between biology and material design can be harnessed to work with the bacteria to produce unique shape change in response to relative humidity. After investigating the various factors influencing outcomes in our early tests, we were able to evolve the material to a point of basic predictability and effectively design shape changes.

These hygromorphic **prototypes ↗** form the basis of my doctoral studies, 'Biodynamic Architectures – a hygromorphic material evolution', which aims to further develop this simple programmability into more predictable deformations capable of complex and refined shape change. Extensive exploration through material tinkering (Parisi et al., 2017) has allowed me to investigate multiple variables such as inert material type, thickness and resistance patterns, spore state, actuator shape and size, spore **adhesion ↗** and **deposition ↗** methodologies. These investigations have developed fluidly through an iterative design process seeking the materials' limits in the sense of reproducible force output and shape change. This has been designed with accessible fabrication methods in mind and applicability to architecture in terms of resilience, feasibility and scale. By asking ourselves these questions when developing a new smart **biomaterial ↗**, it focuses the design methodology on working within the limits of biological capabilities.

Some of the latest developments with this prototype | Figure 4 ↗ | show how altering the surface of the latex substrate to include directional etching can predetermine what deformation shape is achieved in response to relative humidity. This highlights how, with a careful balance of design and biological understanding, the hygromorphic properties of the spores can be harnessed predictably and successfully. Furthermore, by only applying spores to the etched areas, directional 'folds' can be achieved | Figure 5 ↗ |. This discovery opens up the potential of this material for more specific, complex shape change, even reaching the realms of self-assembly | Figure 6 ↗ |.

## Conclusions, Lessons and Future Potential

This bio-smart material shows considerable potential for a benchtop prototype, but it is important to note that there are still numerous hurdles to be overcome to allow transition to the architectural scale. During the development of this research we have created a biodesigners' toolkit to navigate this bio-smart material, the potential of which we are only beginning to realise. Development of complementary technologies in 3D inert material printing would allow progression of the prototype from 3D to 4D motion with more complex architectures. We hope to augment the materials' potential through biomimetic designs while aggregating elements to create more complex interaction and overall response between component parts. The destination field of this material is still relatively unclear and we foresee overlaps into fields such as soft robotics and self-assembly.

This work is part of an ongoing PhD thesis.

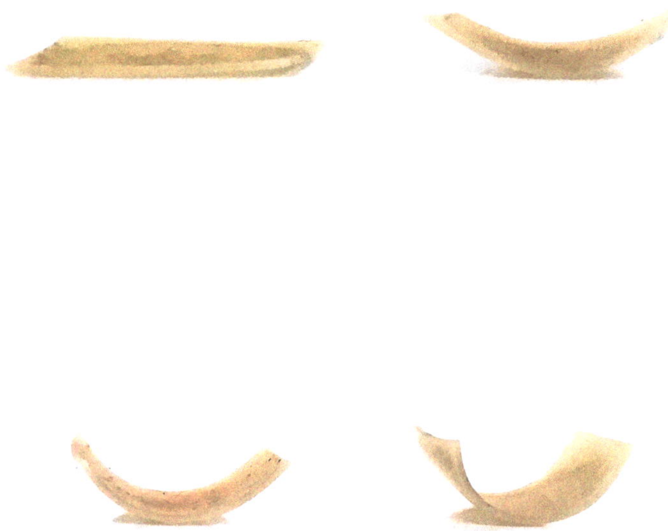

Figure 3

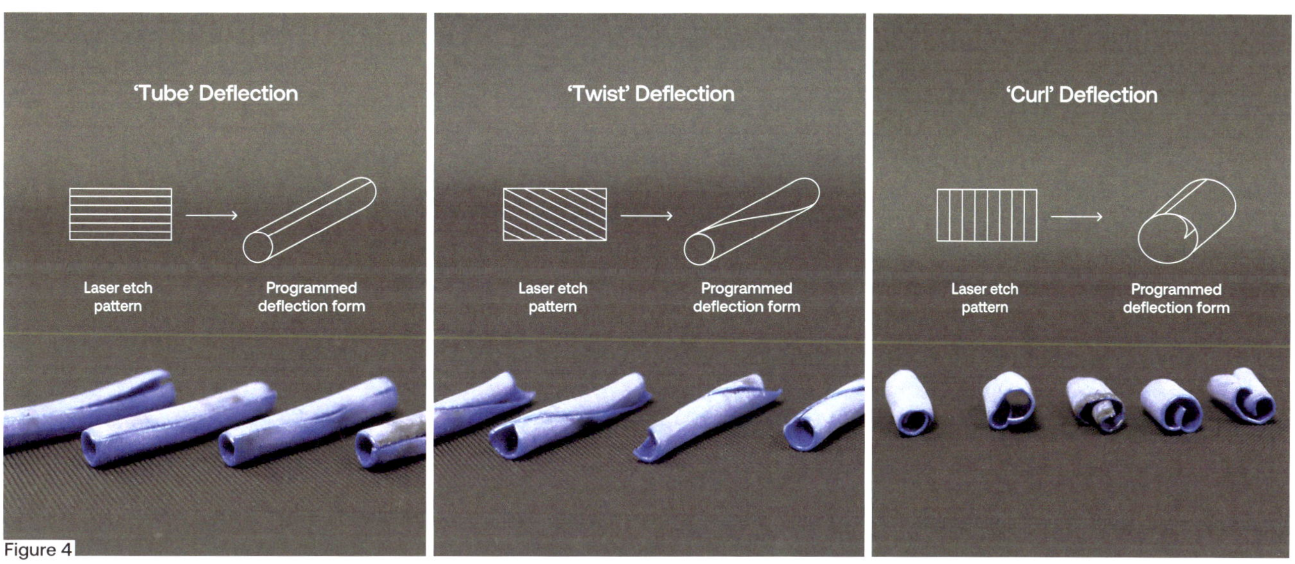


Figure 4

**Figure 3**
Increasing spore concentration increases deflection angle for a given relative humidity

**Figure 4**
Initial results from directional programming - by laser etching different striped patterns into the surface of the latex substrate the deflection outcome can be predetermined

Figure 5

**Figure 5**
Investigations into folding and self-assembly

**Figure 6**
Preliminary exploration into self-assembly

Figure 6

## References


Birch, E., Bridgens, B., Zhang, M., Dade-Robertson, M. (2021) 'Bacterial spore-based hygromorphs: A novel active material with potential for architectural applications', *Sustainability* (Switzerland), 13(7), 1–19.

Koyaz, M. (2017) 'Adaptability Level of Façade Systems Regarding Façade Adaptability Level of Façade Systems'. *Proceedings of Interdisciplinary Perspective for Future Building Envelopes*, Istanbul, Turkey, 15th–18th May 2017.

Lendlein, A., Balk, M., Tarazona, N. A., Gould, O. E. C. (2019) 'Bioperspectives for Shape-Memory Polymers as Shape Programmable, Active Materials', *Biomacromolecules*, 20(10), 3627–3640.

Parisi, S., Rognoli, V., and Sonneveld, M. (2017). 'Material Tinkering. An inspirational approach for experiential learning and envisioning in product design education', *The Design Journal*, 20 (sup 1), S1167–S1184.

# 5.3

Assia Crawford

# 3.3

# Photosynthetic Biocomposites
## Living Microalgae in Minimal Moisture Environments

### Context

↗ **Glossary**
biofilms

Our planet is a unique oasis in the cosmos, sustaining a diversity of life forms. The process responsible for shaping it, photosynthesis, occurs in plants and algae. It is a function present in large terrestrial flora; however, it originated within cyanobacteria, and later in other types of phytoplankton such as microalgae – single-celled photosynthetic organisms typically located within marine environments (Johnson, 2016). These microbial life forms are present within most environments on Earth and have adapted to survive through species and strain diversity | Figure 1 ↗ |. Algae also act as carbon sinks and are believed to be responsible for over 50 per cent of the planet's oxygen production (Stefanova, 2020a). Recent research has been investigating ways of cultivating photosynthetic life within low moisture settings. This has the capacity to open up a new set of design opportunities by integrating densely populated living **biofilms** ↗ within the built environment with higher rates of $CO_2$ sequestration.

**Figure 1**
The algae culture room, incubating various algae and cyanobacteria strains

## Biocomposites and Experimental Fabrication

↗ **Glossary**
biocomposites
hydrogels
kappa-carrageenan
biomaterial

The photosynthetic **biocomposites ↗** highlighted here comprise living microalgae sustained within minimal moisture environments. By developing algae-laden **hydrogels ↗** | Figures 2, 3 ↗ |, we can integrate living materials within various applications that benefit from light exposure, sustained compatible temperatures and humidity levels (Crawford, 2022).

Algae and cyanobacteria do not necessarily exhibit visible changes during their life cycle that can be easily correlated with their health and development (as is the case with other living organisms such as mycelium and bacterial cellulose). This necessitates the use of alternative methodologies for developmental assessment, such as either measuring carbon dioxide sequestration rates directly or inferring the rate of sequestration through other methods such as Pulse Amplitude Modulation Fluorometry (PAM) (Murchie, 2013; In-na, 2020). PAM is a method of light excitation of photosynthetic cells that measures the performance of the photosynthetic cells through qualitative and quantitative data. The latter is of particular interest when assessing microecologies created by designed geometries. Visual readings record differences in algae development along various types of surfaces through non-destructive means | Figure 4 ↗ |.

As we develop various types of biocomposites, there is a particular interest in the interaction between the living and the inanimate. These finely tuned microbial communities are acutely sensitive to their microenvironments, and as such it is essential to establish methodologies for assessing these interactions. This becomes an exercise in establishing base parameters and becoming intimately acquainted with species that reside on a scale often divorced from the visible or the architectural. This presents opportunities to reimagine the process of making by reappropriating traditional techniques, as is the case with ceramic biocomposites | Figure 5 ↗ | that rely on the natural capillary action present within unglazed ceramic structures | Figure 6 ↗ |.

We might also subvert emerging digital tools to serve a purpose beyond commercially prescribed functions. An example of this is the use of pressure-based extrusion to fabricate with bespoke algae-laden matrices, where we employed **kappa-carrageenan ↗** and commercial binders to generate non-toxic, minimal moisture, extrusion-compatible matrices suitable for 3D printing | Figure 7 ↗ | (Stefanova, 2020a). Our work went beyond extrusion and evaluated the potential of the encapsulated microalgae to be sustained for multiple weeks at a time | Figure 8 ↗ |. Preserving the matrices in a dry environment harnesses the natural ability of certain microorganisms to remain dormant in a dehydrated state. This presents the possibility of prefabricating living surfaces and activating them at a later date, making such **biomaterial ↗** solutions viable within current production contexts (Stefanova, 2021).

## 'Soft and Furry' Futures

Photosynthetic biocomposites illustrate speculative thinking and benchtop testing that points to interior applications within buildings. Photosynthetic biocomposites not only bring forth the promise of lower carbon emissions, but of carbon-negative design. These emerging research investigations both challenge traditional views of inanimate and static architecture, and propose new avenues towards an architecture endowed with metabolic functions.

This work is part of a PhD thesis titled: 'Living Building Practice: Designing for a Post-Anthropocene Era'.

Figure 2

Figure 3

**Figure 2**
Coating earthenware ceramic with a living microalgae binder-based hydrogel matrix

**Figure 3**
Testing various clay binder and hydrogel coatings containing living algae cells

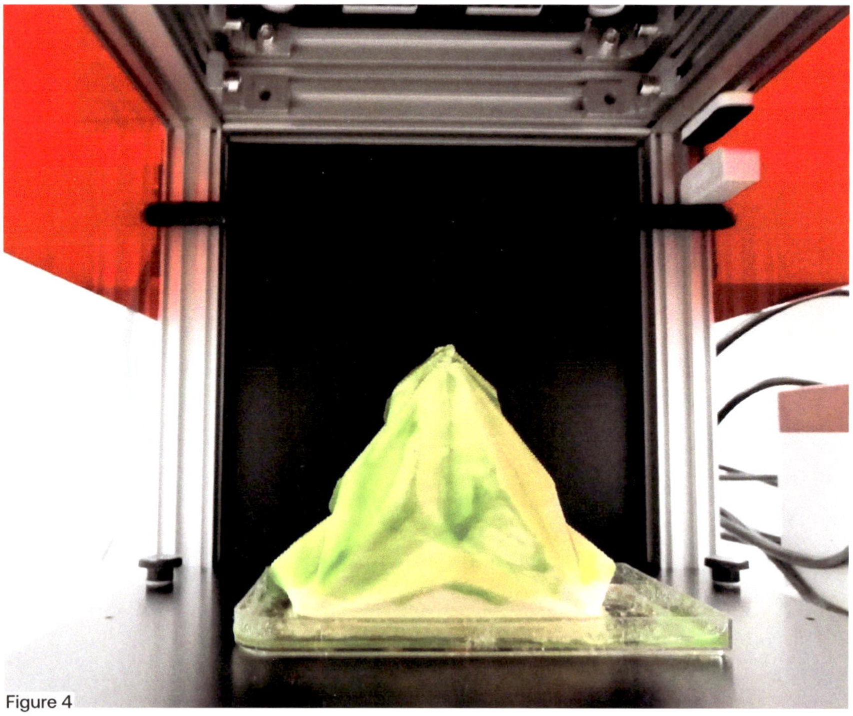
Figure 4

Figure 5

**Figure 4**
Living photosynthetic sample ready for non-destructive reading collection within I-PAM excitation chamber

**Figure 5**
3D printed geometry demonstrating the relationship between complexity and print fidelity

**Figure 6**
3D printing an interlocking building component size ceramic vessel to form as part of a photosynthetic wall with nutrient storage and distribution capabilities

Figure 6

Figure 7

**Figure 7**
3D printed living matrices on various types of textiles, incubated within petri dishes

**Figure 8**
3D printed hollow ceramic vessels coated in a microalgae hydrogel matrix, dried post cultivation

Figure 8

## References


Crawford, A., In-na, P., Caldwell, G., Armstrong, R., Bridgens, B. (2022) 'Clay 3D printing as a bio-design research tool: development of photosynthetic living building components', *Architectural Science Review*, Apr 8, 1–11.

In-na, P., Umar, A. A., Wallace, A. D., Flickinger, M. C., Caldwell, G. S., Lee, J. G. M. (2020) 'Loofah-based microalgae and cyanobacteria biocomposites for intensifying carbon dioxide capture', *Journal of $CO_2$ Utilization*, Dec, 42.

Johnson, M.P. (2016) 'An overview of photosynthesis', *Essays in Biochemistry*, 60(3), 255–73.

Murchie, E. H., and Lawson, T. 'Chlorophyll fluorescence analysis: a guide to good practice and understanding some new applications', *Journal of Experimental Botany*. 2013; 64(13), 3983–98.

Stefanova, A., Bridgens, B., Armstrong, R., In-Na, P., Caldwell, G. S. (2020a) 'Engineering a living building realm: development of protective coatings for photosynthetic ceramic biocomposite materials', The 7th International Conference on Architecture and Built Environment with Architecture Awards. Tokyo: Get It Published Verlag; 362–72.

Stefanova A., Bridgens, B., In-na, P., Caldwell, G., Armstrong, R. (2020b) 'Architectural Laboratory Practice for the Development of Clay and Ceramic-Based Photosynthetic Biocomposites', *Technology | Architecture and Design*, Jul 2;4(2), 200–10.

Stefanova, A., In-na, P., Caldwell, G. S., Bridgens, B., Armstrong, R. (2021) 'Photosynthetic textile biocomposites: Using laboratory testing and digital fabrication to develop flexible living building materials', *Science and Engineering of Composite Materials*, May 20; 28(1), 223–36.

# 3.4

Dilan Ozkan

# 3.4

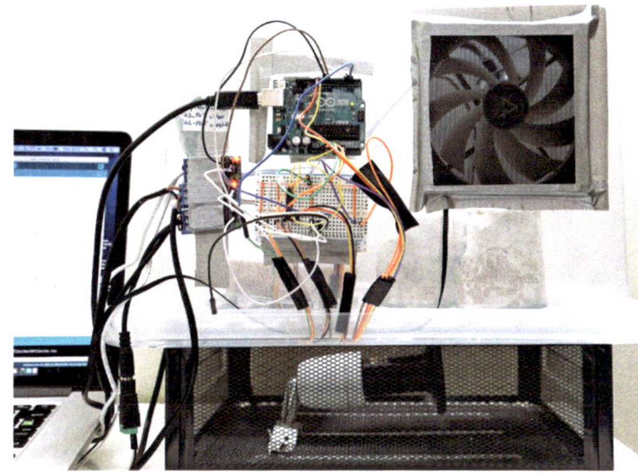

**Figure 2**
Digitally controlled growth chamber which allows for modulation of aspects of the environment such as $CO_2$ and humidity levels

# Designing Mushrooms
## Designing a Living Material Through Bio-digital Fabrication

### Context

↗ **Glossary**
biomaterial probe

There is growing interest in designing with nature where living organisms such as algae, bacteria, and fungi are starting to be used as architectural materials. Unlike traditional building materials, living cells exhibit their own agency: an ability to sense, compute and respond to environmental stimuli. However, their complex biosystems and behaviours are not always predictable and controllable, making them difficult to work with. If we are to work successfully with these materials, we need to explore new methods and processes of fabrication that work alongside their biological processes. Working with digital systems and parameters provides a level of predictability and an understanding of a means of navigating biological complexity.

The research described here focuses on the digital fabrication of living materials. Mycelium (the root system of fungi) can be used to grow bulk materials; however, current approaches tend to be cultivated in moulds. This study suggests a fabrication process that uses fungi as a **biomaterial probe** ↗, in which the fruiting body is grown with minimal

**Figure 1**
Microscopic image of fungal pinheads captured using Dino-Lite digital microscope at 70X magnification

physical intervention but where the growth of the fruiting body is guided through the robotic control of the environment temperature, humidity, $CO_2$ level, and light exposure | Figure 2 ↗ |. The fabrication process uses the natural self-assembly capability of the fungus whilst guiding mushroom growth with parameters as a form of bio-digital fabrication. This method enables the fungus to reveal its inherent potential, which comes from having a metabolism and an ability to grow and adapt autonomously.

Rather than building a catalogue of mushrooms by 3D scanning them | Figure 3 ↗ |, experiments were designed to understand the influence of environmental conditions as parameters, allowing the designer to test the concept of biological parametrics. This questions how far the designer can push the morphology of an organism by setting a range of particular environmental factors. Further, it questions whether parametrics can be used as a method in a predictive way and as a template for other bio-designers working with different organisms. This approach is defined as the Probability Space Method (PSM).

## Parametric Design – Issues and Lessons

↗ **Glossary**
parametric

While testing this **parametric ↗** design approach on fungi, the main issue was the highly complex and non-linear behaviour of living materials. The same effect did not always cause the same morphological results on mushrooms growing under the same environmental conditions. Moreover, in some of the samples, small changes in environmental conditions created tipping points and led to developmental outcomes that were not easy to attribute to single or limited sets of parameters. In essence, it is the non-linear behavioural pattern of the living materials that leads to an abundance of variations in the final product (Ozkan et al., 2022). However, the developmental plasticity of mushrooms allowed me to demonstrate distinct trends and a linear parametric behaviour within the organism. | Figures 4 and 5 ↗ | demonstrate different mushroom morphologies achieved by altering environmental conditions. There was a correlation between environmental parameters and morphological outcomes. What is true for mushrooms may also be true for other sorts of biological systems, and thus a parametric approach to biodigital fabrication can be a new way of crafting living materials.

## Conclusions and Future Potential

The correlation between growth parameters and material response means it is possible to predict the behaviours of biological material in between tipping points. This enables me to work with organisms as they grow in order to achieve the desired form and properties that exist within their plasticity. Therefore, a prediction method or model is needed for non-linear materials that can suggest linearities under given conditions.

Due to the nature of living materials, designers' intent and outcomes do not always proceed in parallel. The Probability Space Method (PSM) is therefore proposed as a way to indicate zones where the growth possibility of the organism is low or high. It is an architectural representation of a spatial range in which the newly grown mushroom can take form. PSM creates a language that everyone can understand, aims to help in reducing uncertainty, and shows how far a designer can step back. It also differs from tolerance. Both in engineering and biology, tolerance is used to determine the amount of space required for error or survival. High to low extreme values determine the range. Probability Space does not represent the range of tolerance; rather, it represents the possible states of the organism for a specific environmental condition | Figure 6 ↗ |.

**Figure 3**
3D Scanning mushroom morphologies using EinScan to record them digitally

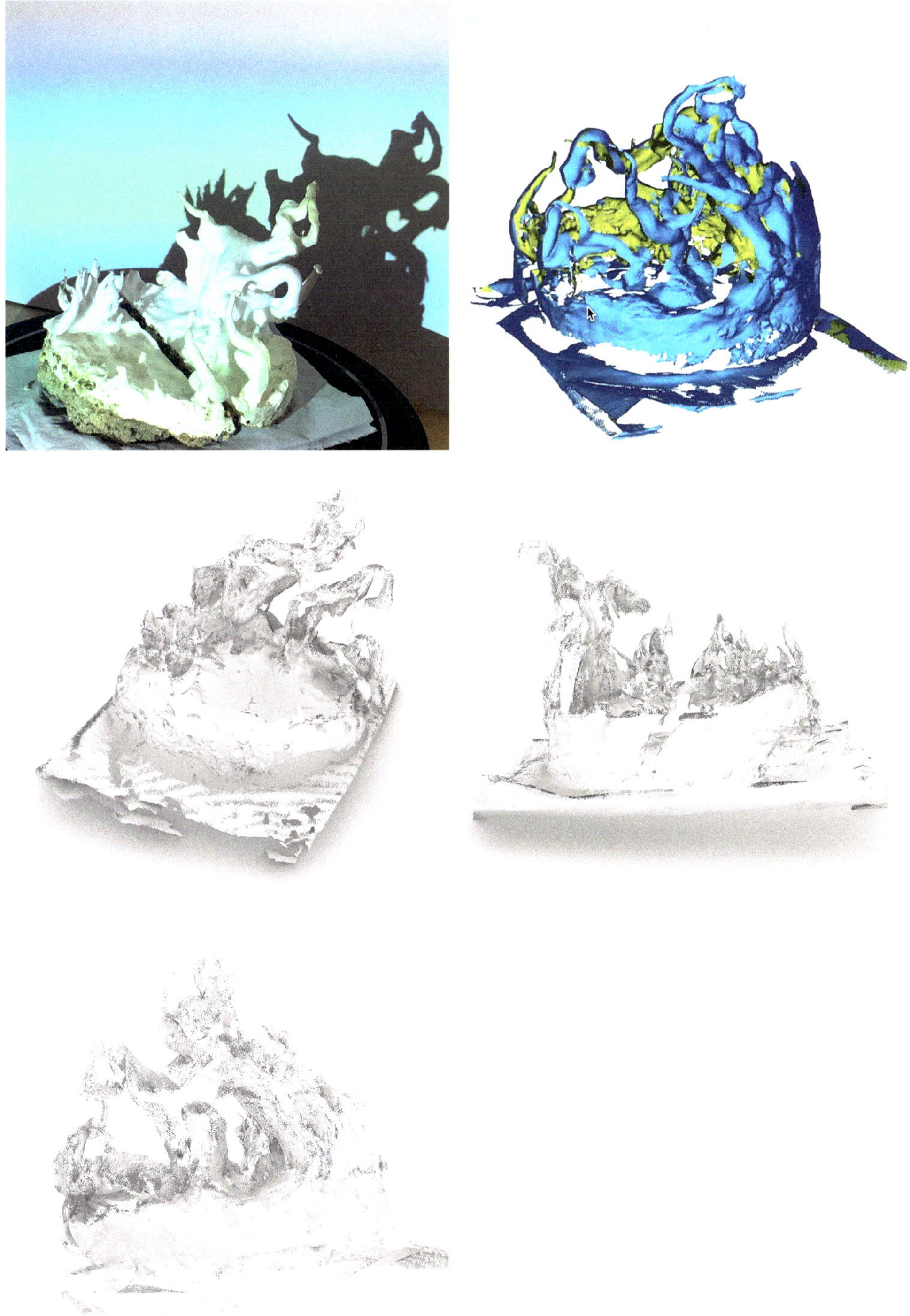

## References

Ozkan, D., Dade-Robertson, M., Morrow, R., Zhang, M., (2022) 'Are mushrooms parametric?' *Biomimetics*, 7(2), 60. Special Issue: Fungal Architectures.

Figure 4

**Figure 4**
Oyster mushroom grown in a higher and lower concentration of carbon dioxide

**Figure 5**
Oyster mushroom grown in high and low humidity

**Figure 6**
The probability space of newly grown mushrooms

Figure 5

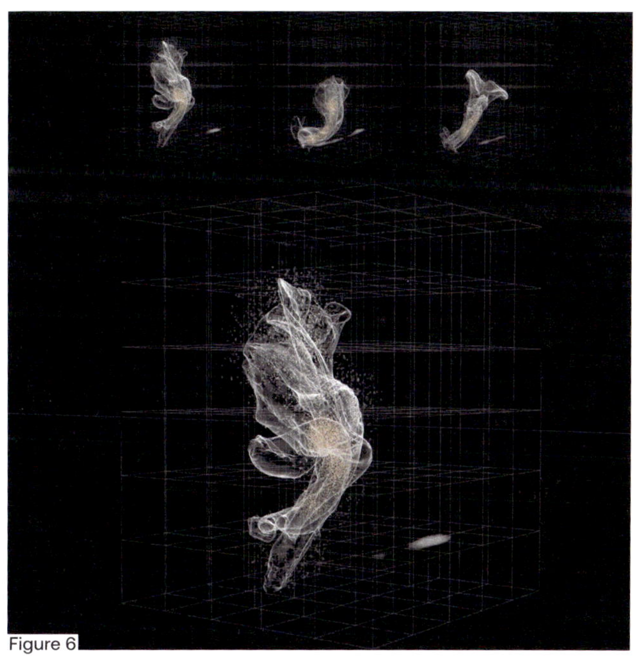

Figure 6

# 3.5

John Allan

# 3.5

# Tiny Urban BioReactor
## Transforming Domestic Waste

### Introduction

↗ **Glossary**
bioreactors
lignin

Deployment of futuristic biotechnologies within home-like settings would permit decentralised bioprocessing to handle domestic waste. Biotechnology can remodel these materials to become useful molecules for the home; for example drugs, flavourings or fuels. Individual homes could have tiny, urban **bioreactors** ↗, where residents can decide how to upgrade their waste materials in ways that best serve them.

A Tiny Urban BioReactor (TUBR) represents an accessible and useful deployment of this approach to biotechnology. It gives people agency over how their waste is handled, and empowers individuals and communities to make useful products. Imagine if the bin at your back gate was producing valuable materials for your home. Paper bins may be replaced with a microbial culture fed by yesterday's newspaper that can generate medicines. Or maybe you'd prefer to utilise a genetically modified strain that can turn **lignin** ↗ into capsaicin (the active component of chilli peppers) to spice up your meals?

**Figure 1**
Use of microbial cultures to convert domestic waste into useful products; in this case making beer from waste paper.

## Unlocking Waste's Potential With Metabolic Engineering

↗ **Glossary**
biomass
metabolites

Central carbon metabolism is the name given to the network of metabolic processes and cycles in living organisms which converts food sources into energy and new **biomass** ↗. As carbon-based organisms, this is normally a process of bursting and reconfiguring the carbon sources which we eat into proteins, DNA and other chemical essentials.

Metabolic engineers genetically modify microbes to drive novel carbon sources into their central carbon metabolism, or draw from it to produce useful **metabolites** ↗. An example of this is the production of opiates from glucose using brewer's yeast (Galanie et al., 2015,). Glucose is a common carbon source in laboratory investigations; however, it is not readily available as a refined, pure feedstock. The use of other carbon sources opens up the potential to build biomass out of abundant, cheap feedstocks; for example household waste.

Paper and plastic materials are essentially different configurations of carbon atoms in long chains with oxygen and hydrogen. While we think of plastics as being an entirely human invention, bacteria have made and degraded their own plastics for millennia. Furthermore, they can degrade plastics found in plants, and are evolving to degrade synthetic plastics.

These waste feedstocks are so abundant that we treat them as a problem. Yet microbes have already adapted to benefit from them. Deeper understanding of the ways they do this allows us to learn how to work *with* them, building microbial partnerships for the bioproduction of critical products that benefit communities in bespoke ways.

## Working With Microbes for Bioprocesses in the Service of People

Violently diverting the metabolism of microbes to utilise novel feedstocks is rarely a sensible strategy. It is akin to driving more work onto one burnt-out team member, where another has abundant time and expertise for it. Researchers now appreciate the deleterious effects of imposing additional genetic programs on the host organism's physiological capacity, and techniques are emerging to respond to this realisation where synthetic processes are minimised when cells are struggling (Ceroni et al., 2018). One way to relieve this burden is to divide labour within communities by designing synthetic microbial consortia and enriching existing ones.

Typically though, the application of these technologies requires bespoke infrastructure and purpose-built facilities. This abstracts the connection people may have to biotechnologies, and robs them of agency regarding how these materials might be used. This begs the question that if people had the option, might they view and utilise their waste differently? Perhaps more responsibly?

## Building a TUBR

The technology to develop these decentralised bioprocesses exists, though there are outstanding questions about how policymakers can handle its effective deployment. However, in practical terms there are few barriers.

A wave of 'community biotechnology' is making a widespread uptake of decentralised bioprocessing not just feasible, but a realistic prediction. Bioreactors may be as simple as home-brew fermentation bins. Demonstrating this, we have utilised paper-digesting bacteria to release sugars from paper waste. The resultant sugar was used in an augmented brewing approach to produce beer.

More complex bioreactors are becoming increasingly accessible to hobbyists and scientists alike. The Chi.Bio platform (Steel et al., 2020) is an example with a reasonable price point. Its open-source availability presents the opportunity to DIY-build modifications or innovate, drawing on its software and concept for new applications.

New ideas in synthetic biology are moving towards harmonising synthetic genetic programs with microbial goals. Decentralised bioprocessing might best be realised by collaboration with community stakeholders holding the waste feedstocks, and community bio groups with an interest in converting it for the public good.

## References


Ceroni, F., Boo, A., Furini, S., Gorochowski, T. E., Borkowski, O., Ladak, Y. N., Awan, A. R., Gilbert, C., Stan, G-B., and Ellis, T. (2018) 'Burden-driven feedback control of gene expression', *Nature Methods* 15(5): 387–393.

Galanie, S., Thodey, K., Trenchard, I. J., Filsinger Interrante, M., and Smolke, C. D. (2015) 'Complete biosynthesis of opioids in yeast', Science 349, 1095–1100.

Steel, H., Habgood R., Kelly C. L., and Papachristodoulou, A. (2020) 'In situ characterisation and manipulation of biological systems with Chi.Bio', *PLOS Biology* 18(7).

Walker, J. T., Strawhacker, A., Angleton, C., Allan, J., Konwar, A., Obayomi, O., and Kong, D.S. (eds.) (2021) 'Proceedings of the Global Community Bio Summit 4.0'.

Ruth Morrow
Ben Bridgens
Louise Mackenzie

# Conclusion
## Branching Out

Bioprotopia is a concept. It is a vision that branches out into many possible architectural futures, each with biology at its heart, prepared to evolve and adapt to the fluid interactions of humans, micro- and macro-organisms, environment and climate. Bioprotopia is also a practical experiment in what is required to bring the emerging field of Biological Architecture to life beyond the confines of the laboratory. **Part 1** explored the social, ethical and political challenges that emerge when we attempt to define the living environment in architectural terms, and the prototypes in **Parts 2** and **3** tested and demonstrated some of the possibilities and practicalities of bringing Biological Architecture into real-world contexts. Thus this book represents the first steps in preparing the ground with hopeful experimentation. Some of these steps may fail to propagate, whilst others may take root and unfurl, developing into a speculative architectural 'Bioprotopia'.

**Figure 1**
The BioKnit prototype behind the glazed façade of the OME

## Bedding In

There are two possible routes to the implementation of biotechnologies in the built environment. The first involves substitutional strategies; for example, using mycelium to make insulation materials (Biohm, 2015), or bio-cement in the making of blocks (Biomason, 2022). There is also the potential for hybrid approaches where biomaterials and bioprocesses are combined with both conventional and/or waste construction materials and methods (Chapter 2.2, Biosandstone). Such strategies may enable implementation within existing regulatory frameworks and design and construction cultures, providing a more rapid route to demonstrate the viability of biotechnologies in the built environment. However, these strategies essentially substitute biological approaches for normative ways of working, and thus sustain current production and manufacturing systems.

The second route involves a radical reimagining of how we construct and inhabit the 'built' environment. This entails engaging at a much deeper level with the more-than-human agents we seek to collaborate with, allowing them to express their preferred forms, rhythms and flows. To enable this radical shift we also need to break out from the norms and regulations that are currently in place in the construction industry – norms in which unsustainable construction practices are deeply embedded. Notably, the Federal Chamber of German Architects recently called for more experimental, and less regulation-driven, methods of construction – known as Part E (BAK, 2022). This is the beginning of a movement that recognises the need for a space – not beyond but within the current building regulations – where architects and engineers can explore new materials, systems and forms of construction.

## Seeking New Paths

There is a tendency to equate the term 'bio' with sustainability. For instance, the description of biofuels as carbon neutral, and therefore sustainable, does not stand up to close inspection. The use of edible crops as first-generation biofuels resulted in unintended and unsustainable consequences that included emissions due to agricultural production, transport and processing; food shortages and hence increased food prices; and environmental damage caused by land use change such as deforestation (Sheehan, 2009).

For biotechnology to enable a shift to a sustainable or regenerative built environment, it is vital that the impacts of 'bio-at-scale' production are acknowledged and critically scrutinised at every stage of development. This will require humility – where promising biotechnological developments may have to be altered or abandoned once the impacts are mapped. It will not be acceptable to develop new biological materials if the processes required to make them use vast quantities of energy and resources. Similarly, it will not be sufficient to substitute one material for another if we continue to store, ship and transport them around the globe in the same damaging ways that we have to date. Such work necessitates even greater interdisciplinary collaboration to embed Life Cycle Assessment and Industrial Ecology expertise into already diverse teams of bioscientists, engineers and architects. For example, it is common for mycelium to be grown using 'agricultural waste', but are these materials really 'waste'? Would they have been beneficially recycled through existing agricultural practice? And what would the impacts be if such 'waste' materials were used at the scale of mass housing? Beyond the painstaking quantification of impacts, creativity, systems thinking and an ethical radar are required in order to anticipate unintended consequences of new production processes, new materials and new systems.

## Collaborative Cultures

We have used prototypes to explore the challenges and opportunities of developing and implementing biotechnologies for the built environment, primarily with a technical, scientific and architectural focus. However, to move towards Bioprotopia, we need to imagine space and place in ways that we have not yet dreamed of. It is likely that such a world will not resemble the orthogonal, homogeneous contemporary architecture and product design that we have become accustomed to in western culture. Rather, through embracing a more-than-human, moist and messy bio-aesthetic, we can realise the potential to transition to a regenerative built environment, where humans are a part of complex self-sustaining ecosystems.

The concept of Bioprotopia also relies on extending our source of precedents and references beyond the scale of the building to the microbial; and beyond materials, processes and practices of predictability, constancy and control to examples of self-determinacy, change and fluidity. This amounts to a new and extensive set of reference points with inevitable contradictions and epistemological tensions. Managing the scale and complexity of these reference points then becomes both a task and an opportunity.

The vision of Bioprotopia therefore cannot be realised without the inclusion of multiple voices. It requires broad cultural engagement that spans industry, policy makers and communities. The future projects of biological architecture must necessarily include the voices of communities who can bring the real world into contact with the laboratory. Marginalised and diverse voices will bring much-needed experience and knowledge, along with the expertise of DIY and hobby communities, ethical and environmental groups, food poverty groups and countless others whose experience of working and living with flora, fauna and microbiota will be invaluable in guiding the course of future research. Further, it will be the role of biological architecture to work with policy-makers and industry to identify and recognise how the processes, systems and materials of biology and biotechnology can provide long-term value on a planetary scale, and not simply alleviate short-term human needs and goals.

Put simply, the future of biological architecture is in our hands, but it will be through the collective intelligence of many peoples and species that Bioprotopia will reveal itself.

## References

BAK, Bundes Architekten Kammer. (2022) https://bak.de/presse/pressemitteilungen/bundeskammerversammlung-verabschiedet-erklaerung-fuer-mehr-spielraum-und-innovation-beim-planen-und-bauen/ (Accessed: 30 Nov 2022).

Biohm. (2015) https://www.biohm.co.uk/mycelium (Accessed: 30 Nov 2022).

Biomason. (2022) https://biomason.com/ (Accessed: 30 Nov 2022).

Sheehan, J. J. (2009) 'Biofuels and the conundrum of sustainability', *Current Opinion in Biotechnology*, 20(3), 318–324.

# Biographies

Angie Sherry
Assia Crawford
Ben Bridgens
Dilan Ozkan
Emily Birch
Jane Scott
John Allan
Louise Mackenzie
Magdalini Theodoridou
Peg Rawes
Ruth Morrow
Thora Arnardottir

**Angie Sherry** is a Vice Chancellor's Senior Fellow in Environmental Molecular Microbiology, HBBE, Northumbria University, UK. Her research aims to mechanistically understand the function and composition of microbes in natural and built environments, including impacts associated with environmental change or engineering interventions. She uses a range of traditional microbiology methods together with high-throughput 'omics technologies to decipher environmental microbiomes; working on collaborative ventures with industry and academia in interdisciplinary teams. Her current research includes built environment microbiomes, fungal/fibre highways towards enhanced oil bioremediation and carbonate precipitation in microbial mats, masonry, concrete and engineered soils.

**Assia Crawford** is an Assistant Professor in Technology at the College of Architecture and Planning, University of Colorado Denver, whose creative practice research focuses on the development of biological material alternatives and digital fabrication practices for a post-Anthropocene era. Her work sits on the intersection between architecture, science and critical theory and employs experimental and speculative design to address ecological challenges faced by communities at a time of environmental uncertainty. Assia is an ARB registered architect and is the recipient of ACSA TAD Best Article Award (Volume 4) for 'Architectural Laboratory Practice for the Development of Clay and Ceramic-Based Photosynthetic Biocomposites'.

**Ben Bridgens** is a Senior Lecturer in Architectural Technology in the School of Architecture, Planning & Landscape at Newcastle University, UK, and a founding member of the HBBE. Ben works at the interface of structural engineering, architecture and design, critically examining 'sustainable' (bio)technologies and exploring the potential of reimagining low-tech, traditional approaches for the construction and operation of the built environment. Ben led the design of an experimental house on the Newcastle University campus called 'The OME', and co-leads the HBBE's 'Responsible Interactions' theme.

**Dilan Ozkan** is an architect and researcher who focuses on working with living systems. She aims to bring multidisciplinary and innovative approaches to architectural production by discovering new material systems. Dilan completed an architectural design Master's at Pratt Institute in New York, where she was speculating about the near-future built environment scenarios. Currently, she is a PhD student at Newcastle University and investigating the digital fabrication of non-linear materials over fungi. During her PhD, she formed a study group called Mycology for Architecture to collaborate with other disciplines and share knowledge about fungi.

**Emily Birch** is a multidisciplinary researcher exploring the possibilities of designing and synthesising dynamic biological elements for application within modern architectures. She has a keen focus on innovative, interdisciplinary working. Currently, throughout her PhD studies she has been focusing on understanding the natural processes behind hygromorphic behaviours in bacteria and incorporating them within flexible materials to produce environmentally responsive structures. Prior to starting her PhD studies, she earned a Bachelor's Degree in Architecture from Newcastle University and intends to continue her architecture education alongside her research in the future.

**Jane Scott** is a Newcastle University Academic Track Fellow in HBBE where she leads the Living Textiles Research Group. She is a knit specialist and her interdisciplinary research is located at the interface of programmable textiles, architecture, and biology. Before joining Newcastle, Jane was an academic at the University of Leeds and held a Visiting Research Fellowship in Biomimicry at Central Saint Martins. She completed her doctorate in Programmable Knitting, at the Textiles Futures Research Centre, UAL. Her work is exhibited internationally, and she has presented research at MIT, the Pompidou Centre, the Design Museum, and the Microsoft Research Centre, Cambridge.

**John Allan** is a microbial synthetic biologist focused on developing contemporary biotech approaches which work in harmony with microbes. Brought to the synthetic biology field through iGEM, John has continued to build an interdisciplinary attitude to his work. He has previously worked on instigating physical sensory mechanisms in bacteria, developing novel bespoke chassis organisms, and designer enzymatic machinery. Currently John is a postdoctoral researcher in Engineering Science at the University of Oxford, building bio-robotic interactions for the control of microbial communities and bioprocesses.

**Louise Mackenzie** is an artist, curator and research associate at HBBE. Her practice explores human relationships with the non-human world and her research has developed a collaborative, interdisciplinary approach to critical engagement with scientific practice. Louise holds a PhD in Fine Art from BxNU Institute of Contemporary Art and is a Director of ASCUS Art and Science in Edinburgh. She is co-founder and curator of an interdisciplinary cinema project, Black Box, and co-founder of Alive Together, an international community for research in human/animal relationships. Her work has been exhibited nationally and internationally, including ZKM (Germany), BALTIC CCA (UK) and National Library of Madrid (Spain), and she has written for MIT Press, Routledge, Bloomsbury and Intellect.

**Magdalini Theodoridou** is a Newcastle University Academic Track Fellow in HBBE and School of Engineering, leading research on novel biotechnological processes and applications for sustainable building and smart heritage conservation. She has worked as a research and teaching fellow in Italy, Hungary, Cyprus and the UK. Her scientific contributions to the design and development of novel construction materials, self-healing geological materials and structures, and the innovative application of micro/non-destructive techniques have attracted international funding and individual research fellowship awards, e.g. the EU Marie Skłodowska-Curie Actions. She is a member of the Heritage NUCoRE Executive team, ICOMOS-ISCS and four RILEM technical committees.

**Peg Rawes** is Professor of Architecture and Philosophy at the Bartlett School of Architecture, UCL. Trained in art history and philosophy, her publications on material, political and ecological architectures include: 'Visualising uncertainty and vulnerability', Materia Arquitectura Journal (2021), and anthologies Relational Architectural Ecologies: Architecture, Nature and Subjectivity (2013) and Poetic Biopolitics: Practices of Relation in Architecture and the Arts (2016), which publish architects alongside practitioners in the arts, environmental, human rights, social and medical research.

**Ruth Morrow** is Professor of Biological Architecture. Ruth's research is largely practice-based and encompasses both the material, the social and the ecological. It is driven by an inclusive, feminist ethos and uses tactics of creativity, collaboration and reflection through writing. She has extensive experience in developing material ideas from concept through to commercialisation, resulting in international funding, design awards, exhibitions, chapters, papers, books and citations. She is currently co-head of the interdisciplinary School X at Newcastle University, where she also co-leads the research theme, Responsible Interactions, in the Hub for Biotechnology in the Built Environment.

**Thora Arnardottir** is a researcher and a multidisciplinary designer specialising in biofabrication, a practice that combines design methods, biological systems, and innovative fabrication techniques. She is particularly interested in integrating living processes in the built environment and harnessing the power of unruly matters to establish a collaborative approach with biological entities. After completing her BA in Architecture at the Arts University Bournemouth, she attended the MA program of the Institute for Advanced Architecture of Catalonia (IAAC) in Barcelona, which inspired her to further develop her work on bacterial biomineralisation in her PhD at Newcastle University.

# Acknowledgements

The Editors would like to fully acknowledge the work and support of the HBBE research themes: Building Metabolism led by Gary Black; Living Construction led by Martyn Dade-Robertson and Meng Zhang; Microbial Environments led by Darren Smith; and Responsible Interactions led by Ben Bridgens and Ruth Morrow.

Thanks to all our colleagues and collaborators, past and present, especially Karolina Bloch, Armand Agraviador, Carmen McLeod and Kaajal Modi. We also want to acknowledge the financial and administrative support provided by Northumbria University and Newcastle University, alongside Research England whose funding established the Hub for Biotechnology in the Built Environment and its experimental building, the OME.

# Colophon

**Editors**
Ruth Morrow, Ben Bridgens, Louise Mackenzie

**Acquisitions editor**
Baharak Tajbakhsh

**Project management**
Angelika Gaal, Freya Mohr

**Production**
Angelika Gaal

**Copy editing**
Patricia Kot, Alun Brown

**Image editing**
Pixelstorm Litho & Digital Imaging

**Layout, cover design and typesetting**
HE&AD Büro für Gestaltung

**Printing**
Grafisches Centrum Cuno GmbH & Co. KG

Library of Congress Control Number: 2023933331

Bibliographic information published by the German National Library
The German National Library lists this publication in the Deutsche Nationalbibliografie; detailed bibliographic data are available on the Internet at http://dnb.dnb.de.



ISBN 978-3-0356-2579-0

e-ISBN (PDF) 978-3-0356-2580-6


P.O. Box 44, 4009 Basel, Switzerland
Part of Walter de Gruyter GmbH, Berlin/Boston

9 8 7 6 5 4 3 2 1
www.birkhauser.com

**Image Credits**
Page/Figure

**Armand Agraviador** 25/4, 59/1, 61/2, 64/6, 66/7, 67/8, 77/2, 85/8, 91/2, 96/9, 103/2, 107/6+7, 117/1, 131/1
**Mahab Aljannat** 86/13-15
**Thora Arnardottir** 149/1, 150/1, 152/2, 153/3+4, 155/5
**Emily Birch** 156/1, 157/2, 159/3+4, 160/5, 161/6
**Karolina Bloch** 80/6, 82/9, 83/10, 85/8, 85/11, 87/16, 96/8
**Ben Bridgens** 12/2, 20/1, 23/2, 25/3, 27/5, 28/6+7, 29/8, 30/9, 33/10, 55/1, 79/3, 90/1, 92/3, 96/10, 100/15, 102/1, 104/3, 113/12, 114/13, 116, 124/9, 125/11, 136/8, 180/1
**Roxana Caplan** 141/12
**Assia Crawford** 162/1, 165/2+3, 166/4-6, 168/7, 169/8
**Martyn Dade-Robertson** 12/2
**Rory Doherty** 99/13+14
**Elise Elsacker** 72/13
**Elizabeth Gilligan** 93/4, 95/5-7
**Iulianiya Grigoryeva** 130, 141/11
**Jamie Haystead** 105/4
**HBBE** 71/11, 72/12, 106/5, 109/8+10, 110/9, 113/11, 117/2, 119/3+4, 120/5, 121/6, 122/7, 123/8, 125/10, 127/12, 129/13, 132/2, 133/3, 135/4-6, 136/7, 139/10, 144/14, 171/2
**Aileen Hoenerloh (PGR)** 65/5, 71/10
**Pippa McLeod-Brown** 136/9
**Kaajal Modi** 78/4, 81/7
**Ruth Morrow** 10/1, 76/1, 78/5, 85/8, 84/12, 85/12, 97/11
**Dilan Ozkan** 65/4, 69/9, 170/1, 173/3, 174/4, 175/5+6
**Jane Scott** 63/3
**Angela Sherry** 98/12
**Ollie Spurr** 142/13
**Ahmet Topcu** 60/1, 72/14, 74/15
**Greg Young** 176/1